PRAISE FOR *THE LIVING SOIL HANDBOOK*

"Over my years practicing no-till market gardening, I've come to truly appreciate listening to *The No-Till Market Garden Podcast* and Farmer Jesse's exploration of no-till systems. Now, this research is inked into Jesse's very well-written and valuable guide, *The Living Soil Handbook*. This book is a gold mine filled with tips, tricks, and effective practices you can apply to your crop itineraries. I advise any grower to follow Jesse's mantra: A no-till system is not a dogma, it's a direction."

—JEAN-MARTIN FORTIER, author of *The Market Gardener*

"The best way to produce healthier soils, fight climate change, and reduce work all at the same time is to disturb the soil less. *The Living Soil Handbook* shows growers how to do just that. I highly recommend this practical and beautifully designed book."

—BEN HARTMAN, author of *The Lean Farm* and *The Lean Farm Guide to Growing Vegetables*

"Jesse Frost's *The Living Soil Handbook* is a terrific, practical application of the no-till principles for which he and his *No-Till Market Garden Podcast* have become known and respected. Disturbing the soil as little as possible—even when managing garden paths, for example—is one theme of this book about letting the living soil live and how to do so. Beautifully illustrated, this is a great read full of useful advice that will perfect your growing game."

—JEFF LOWENFELS, author of *Teaming with Fungi*

"In this wonderful new book, Jesse Frost offers a clear and friendly explanation of why and how you can grow successfully when your methods are fully in tune with nature's processes. Beautifully illustrated by Jesse's wife, Hannah Crabtree, *The Living Soil Handbook* provides a full range of experience-based advice to aspiring growers and gardeners on major topics such as soil fertility and mulches as well as small but important details like bed and path width. Jesse values practicality over dogma, and keeps it achievable: 'Disturb the soil as little as possible.'"

—CHARLES DOWDING, creator of Charles Dowding's No-Dig Gardening Course

"*The Living Soil Handbook* is a must-have resource for those who wish to reduce or eliminate tillage, build soil biology, intensify production, and create a more ecological, regenerative, and successful farm. Farmer Jesse integrates the experiences of a multitude of farmers and his years of research with pertinent soil science in this easy-to-read guide to help grow more resilient farms in the face of climate chaos. It all goes back to the soil and building life!"

—ELIZABETH AND PAUL KAISER, founders and farmers, Singing Frogs Farm

"*The Living Soil Handbook* is a must-read for growers who want to achieve the long-held organic objective of feeding crops by feeding the soil. It goes beyond the mechanics of no-till to explain why it's important to keep the soil 'as undisturbed, as well covered, and as fully planted as possible.' With the understanding of why to do these things, growers can customize soil care systems for any region. Whether or not your goal is to go completely no-till, Jesse Frost's book is a great companion to help you figure out how to 'disturb the soil as little as you possibly can *in your context*.' With an emphasis on understanding soil ecosystems, this book allows growers to improvise their own solutions rooted in soil health."

—ANDREW MEFFERD, editor, *Growing for Market* magazine; author of *The Greenhouse and Hoophouse Grower's Handbook* and *The Organic No-Till Farming Revolution*

"While no-till growing has been popular for amateur gardeners for some time, it is only more recently that commercial growers have embraced its potential. *The Living Soil Handbook* is beautifully clear, making both the complexity of soil biology and the technical crop detail engaging and accessible. Jesse Frost demonstrates the benefits of using no till methods and he also takes us through, in some detail, the range of methods possible at different scales. He is no starry-eyed evangelist though. He explores his failures as well as what has worked well, and points out areas where more research and trials are needed, for instance in successful crop termination. Though this book is aimed at the ecological market gardener, anyone with an interest in growing vegetables with the minimal impact on their soil will thoroughly enjoy and learn from Jesse's sound advice.

—Ben Raskin,
head of horticulture and agroforestry,
Soil Association; author of
The Woodchip Handbook

"As a lifelong farmer who is skeptical of absolute practices and catchphrases like 'no-till,' I'm happy to say that Jesse Frost has done an excellent job of compiling resources and information to explain the tenets of healthy living soil. With a skillful, personable writing style, Jesse offers effective farming techniques and provides a compelling case to disturb the soil as little as possible as well as to keep it planted and covered as much as possible. *The Living Soil Handbook* is a great read for beginning and seasoned farmers alike."

—Clara Coleman,
owner and operator, Four Season Farm;
creator of #RealFarmerCare

"*The Living Soil Handbook* speaks to Jesse Frost's experimental and inquisitive nature whilst seeking out practical and reliable solutions. Garnering wisdom from growers in many regions, as well as from his own experience, Jesse delves deep into what I consider an optimal approach to annual vegetable production. This book explores the pioneering no-dig market gardening system with deep woodchip pathways that I have established at Ridgedale, along with many other complementary approaches for achieving the same outcomes: thriving soil biology, practical workflows, and abundant harvests. It proves once again that it is our pattern-thinking that is important, and that we have a multitude of solutions at our disposal. We are microbe farmers, after all, and this book is a great addition to the literature to help you achieve beautiful and bountiful results."

—Richard Perkins, author of
Regenerative Agriculture and
Ridgedale Farm Builds

"Jesse Frost has made an invaluable addition to the nascent library of no-till market garden manuals. If you want to grow vegetables without tillage, read this book closely and reference it often. Like crops growing from a vibrant soil food web, Jesse's insights pull from interactions with innovative no-till growers across the United States and beyond—and bear fruit worth savoring. Jesse has synthesized this incredible diversity into a comprehensive manual that takes no-till to a deeper level. I learned something new on almost every page. A magnificent union of information gathering and first-person know-how, *The Living Soil Handbook* is a must-read for every soil caretaker."

—Daniel Mays, author of *The No-Till Organic Vegetable Farm*

THE LIVING SOIL HANDBOOK

THE LIVING SOIL HANDBOOK

THE NO-TILL GROWER'S GUIDE TO ECOLOGICAL MARKET GARDENING

JESSE FROST

ILLUSTRATIONS BY HANNAH CRABTREE

Chelsea Green Publishing
White River Junction, Vermont
London, UK

Project Manager: Alexander Bullett
Editor: Fern Marshall Bradley
Copy Editor: Diane Durrett
Proofreader: Lisa Himes
Indexer: Shana Milkie
Designer: Melissa Jacobson

Printed in the United States of America.
First printing July 2021.
10 9 8 7 6 5 4 3 2 1 21 22 23 24 25

Our Commitment to Green Publishing
Chelsea Green sees publishing as a tool for cultural change and ecological stewardship. We strive to align our book manufacturing practices with our editorial mission and to reduce the impact of our business enterprise in the environment. We print our books and catalogs on chlorine-free recycled paper, using vegetable-based inks whenever possible. This book may cost slightly more because it was printed on paper from responsibly managed forests, and we hope you'll agree that it's worth it. *The Living Soil Handbook* was printed on paper supplied by Versa Press that is certified by the Forest Stewardship Council.

Library of Congress Cataloging-in-Publication Data
Names: Frost, Jesse, 1982– author.
Title: The living soil handbook: the no-till grower's guide to ecological market gardening / Jesse Frost.
Description: First printing edition. | White River Junction, Vermont: Chelsea Green Publishing, 2021. | Includes bibliographical references and index.
Identifiers: LCCN 2021012288 (print) | LCCN 2021012289 (ebook) | ISBN 9781645020264 (paperback) | ISBN 9781645020271 (ebook)
Subjects: LCSH: No-tillage—Handbooks, manuals, etc. | Soil management—Handbooks, manuals, etc. | Organic gardening—Handbooks, manuals, etc. | Compost—Handbooks, manuals, etc. | Mulching—Handbooks, manuals, etc. | Soil fertility—Handbooks, manuals, etc.
Classification: LCC S604 .F767 2021 (print) | LCC S604 (ebook) | DDC 631.5/814—dc23
LC record available at https://lccn.loc.gov/2021012288
LC ebook record available at https://lccn.loc.gov/2021012289

Chelsea Green Publishing
85 North Main Street, Suite 120
White River Junction, Vermont USA

Somerset House
London, UK

www.chelseagreen.com

To the incredible moms in my life:
Hannah Crabtree, Lisa Crabtree, Christy Diaz,
Carolyn Breeding, the late Debra Frost,
and Mother Nature (obviously).

CONTENTS

Acknowledgments *xi*

Introduction 1

PART ONE

Disturb as Little as Possible

1: The Basic Science of Living Soil 9

How Photosynthesis Feeds the Soil, 11 • The Five Keys to Photosynthesis, 15 • Defining Tillage, 22

2: Breaking New Ground 33

Site Selection Considerations, 33 • Starting from Scratch, 35 • The Never-Till Approach, 42 • Animal Tillage, 44 • Transitioning to No-Till, 45 • Designing Permanent Beds, 47 • Establishing No-Till Garden Beds, 53

PART TWO

Keep It Covered as Much as Possible

3: Compost in the No-Till Garden 59

The Four Types of Compost, 60 • Risks with Compost, 67 • The Deep Compost Mulch System, 69

4: Mulch 71

Straw, 72 • Hay, 74 • Fresh Hay, Haylage, and Grass Clippings, 77 • Cardboard and Mulch Paper, 78 • Wood Chips, Sawdust, and Bark Mulch, 80 • Leaves and Leaf Mold, 82 • Peat Moss, 83 • Synthetic Mulches, 85 • Cover Crops, 86

5: Turning Over Beds 89
Maintaining Soil Health in a Bed Flip, 91 • The Good and Bad of Occultation, 96 • Weed Whacker, Knife, and Scything Bed Flips, 102 • Mowing Methods, 107 • Stirrup and Wheel Hoes, 110 • Solarization, 111

6: Path Management 113
Wood Chips and Other Mulches, 113 • Plastic Mulches, 119 • Living Pathways, 120 • Mulch-in-Place, 125 • No Mulch in Pathways, 127

PART THREE
Keep It Planted as Much as Possible

7: Fertility Management 131
Measuring and Managing Fertility, 131 • Designing a Fertility Program, 142 • Using Cover Crops for Fertility, 148 • Bed Preparation without Tillage, 159

8: Transplanting and Interplanting 165
Growing Healthy Transplants, 165 • Basic Interplanting Strategies, 169 • Advanced Interplanting Strategies, 180

9: Seven No-Till Crops from Start to Finish 193
Carrots, 195 • Arugula, 201 • Garlic, 205 • Lettuce, 212 • Sweet Potatoes, 220 • Beets, 226 • Cherry Tomatoes, 230 • Some Thoughts in Closing, 239

Appendix A: Cover Crop Use and Termination Guide 241
Appendix B: Critical Period of Competition and Interplant Pairings 247

Resources and Recommended Reading 255
Notes 259
Bibliography 265
Index 273

Acknowledgments

I know where to begin here, but have no idea how. Indeed, I lack the talent as a writer to adequately convey the gratitude I have for my partner, Hannah Crabtree (who is also the illustrator of this book), for all she has done to make the book possible. From taking the kids outside while I conducted interviews to fielding my ideas at all hours—neither this book, nor anything else I do, would exist without her.

Thank you to every farmer and researcher and scientist who lent me their time for interviews. A special thank you to Dr. Aaron Hawkins, Dr. Christine Jones, Dr. Alex Harmon-Threat, Dr. Robert Houtz, Dr. Clarence Swanton, Dr. Bob Hartzler, and Cary Oshins for their extra help with the science. A special thank you also to Jackson Rollet and Josh Sattin for taking up my slack at No-Till Growers (www.notillgrowers.com) while I put this thing together. Thank you as well to Jean-Martin Fortier for his years of support. Thank you to all the amazing Patreon and Venmo supporters. You absolutely rock.

An enormous thank you to Carolyn Breeding, Willie Breeding, and the entire Breeding family for over 25 years of being awesome. Thank you to Cher and Eric Smith for getting me started in farming (and introducing me to Hannah). Thank you to my in-laws Lisa and Joe for all the support and friendship. Thank you to Ellis and Further for being a daily inspiration. Thank you to my father, my mother, Christy, Amanda, Tim, and Pete. And thank you to my editor Fern Marshall Bradley and the good people at Chelsea Green. It's been my dream to write this book, but it would be nothing more than a heap of passive voice and blurry photos without you. Thank you all.

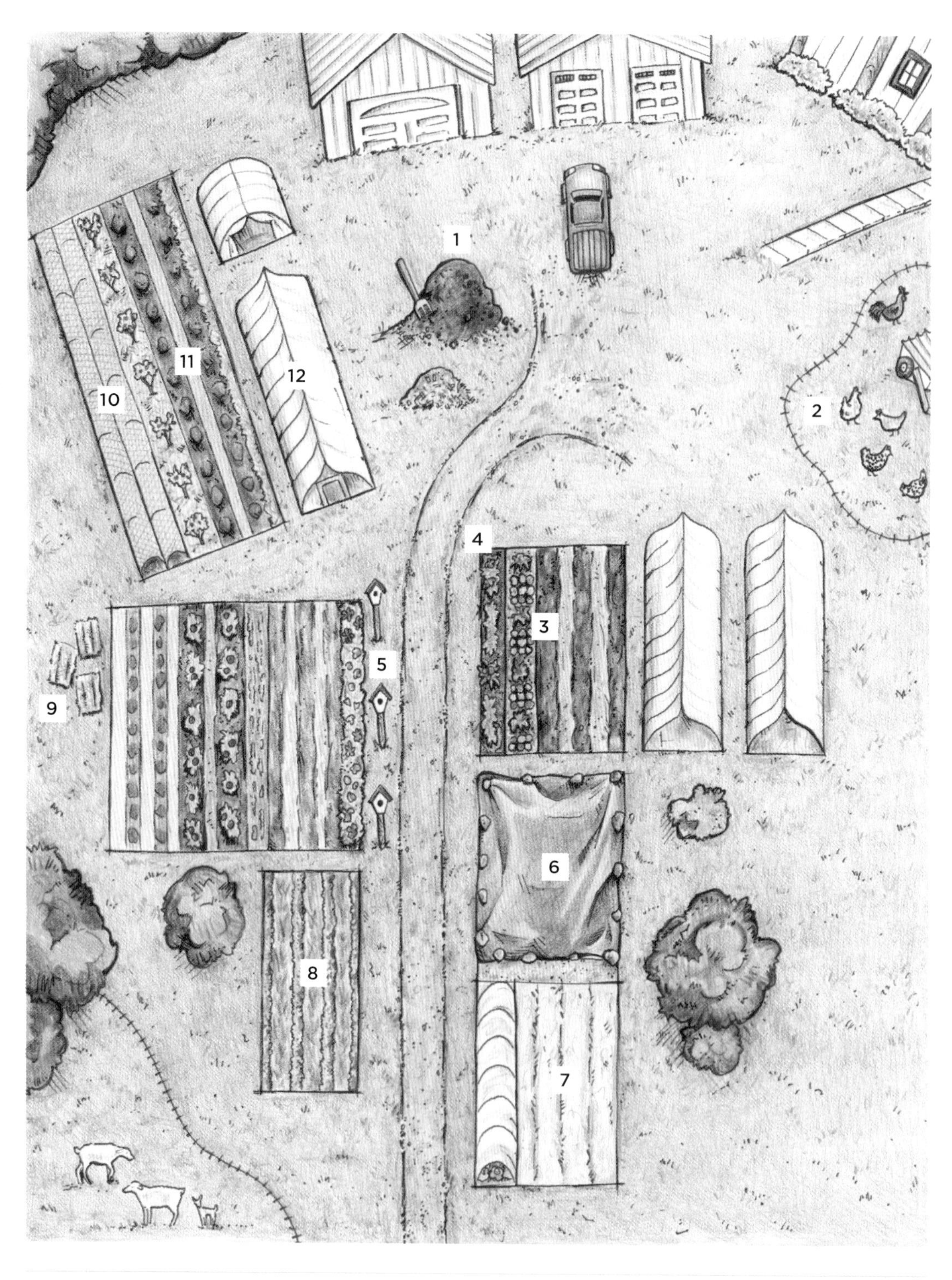

Elements of a No-Till Market Farm. 1. Compost + Wood Chips 2. Animals 3. Interplanting 4. Flowers 5. Hedgerows 6. Tarping 7. Cover crops 8. Living pathways 9. Hay bales 10. Shade cloth 11. Perennials 12. Propagation house

INTRODUCTION

Confession: I have never actually grown anything in my life.

I have never constructed a leaf or imbued a flower with an appealing fragrance to draw in pollinators. I have never sewed roots through soil or traded carbon cocktails with soil microbes in exchange for nutrients. I am just not that cool.

In the 11 years I have been farming, all I can claim credit for is making the conditions right (and sometimes, admittedly, very wrong) for food and flowers to grow. If a customer thanks me for growing the food they purchase, I feel like a fraud. I feel as though I couldn't possibly take that credit. My job—indeed, the job of any grower—is not to grow food but rather to facilitate that growth. Something else entirely does the growing.

That "something" is a complex community of living organisms—both macro and micro—that work in conjunction with air, water, sunlight, carbon, and nutrients to grow plants. Humans aren't the creators here. I repeat: We simply make the conditions right for crops to grow and make food—this is the literal definition of *cultivation*.

Three Principles to Farm By

In this book I blend my experience stewarding living soil with the realities of making a living as a professional grower. The very short version of that knowledge is this: Getting what you need from the soil comes down to first asking the soil what *it* needs. And it is true no matter where you live. What the soil needs to thrive in humid Florida is largely the same as what it needs in dry Montana. It comes down to three basic principles:

1. Disturb the soil as little as possible.
2. Keep the soil covered as much as possible.
3. Keep the soil planted as much as possible.

I first came across these three principles several years ago as a beginning farmer reading about conservation agriculture and soil health. My wife, Hannah, and I were suffering through some crop failures and I sought guidance on what we were doing wrong. The books and articles told me that, although we could apply sprays and try a variety of techniques to protect crops, the *best* way to fight plant disease and pest pressure was to nurture soil health. And the best way to do that? Follow those three principles.

Unfortunately, the books and articles weren't overflowing with guidance on how to follow those three principles. The texts used terms such as *interplanting* and *no-till* or *cover cropping* but did not offer much technical detail on how to execute those practices. Somewhat frustrated, we began experimenting on our farm with eliminating mechanical tillage, trialing different mulches, and interplanting multiple crops together in the same bed to see what liked growing together. In 2018, I started *The No-Till Market Garden Podcast*, and my motive was to help others and myself by having conversations with farmers who were experimenting with low- or no-tillage methods to discover, and then share, what they'd learned. Farmer Jackson Rolett and I started No-Till Growers (www.notillgrowers.com) to aggregate (and create) videos, talks, podcasts, and articles. Later we employed grower Josh Sattin to make detailed technical videos and host a bimonthly live show on YouTube called *Growers Live*. On that show, Sattin interviewed growers, and anyone could log on and ask those growers specific questions.

The goal of all these ventures has been, and is, the same—to answer the question *what does the soil need to thrive?* Ultimately, through these experiences and many conversations with agronomists, growers, and scientists, I've learned about a range of widely applicable technical solutions for keeping the soil as undisturbed, as well covered, and as fully planted as possible. In this book I work to flesh out the details of how to employ those principles not just on a farm like mine, but on any farm. My hope is that anyone, anywhere will be able to use this book as a guide to designing the right system for their context and soil—that is, to put those three principles into practice.

That system might wind up looking similar to the shallow compost mulch system Hannah and I use at Rough Draft Farmstead in central Kentucky (USDA Hardiness Zone 6b) as described in chapter two and throughout this book. Or you may find that some or all of our methods won't work for you. For example, you may not have access to the rich and plentiful compost that we enjoy here in horse country. Furthermore, you might not have the abundant rainfall we do, or the relatively generous number of frost-free days. Environmentally, you might be opposed to the use of plastic silage tarps—and not without reason. To account for that, I've set up this book as a choose-your-own-adventure of sorts. And no doubt, an adventure it will be.

Before I wrap up these introductory thoughts, however, I want to have an obligatory pause and reflect on two crucial words that show up in each of the three guiding principles: "as possible."

Figure 0.1. All of our practices at Rough Draft Farmstead, from mulching to cover cropping to interplanting, are part of our goal to protect and nurture the soil.

Marry. Those. Words.

When the practice of no-till is a grower's primary tool for stewarding the soil, "as possible" must be their mantra. These words are beautifully, even pristinely, the essence of no-till agriculture. They encourage the grower to be reasonable. "Yes," those words remind us, "pulling carrots disturbs the soil. Raking disturbs soil. Animals disturb soil. It's okay. Just disturb the soil as little as you possibly can *in your context.*"

Though avoiding soil disturbance as much as possible is important, the enterprise of creating and protecting living soil isn't beholden to the goal of no disturbance ever. Indeed, I believe each farmer will discover that their path to stewarding living soil evolves as much through dedication to no dogma as it does to no disturbance. As long as you use a given tool to promote soil life and biology, you are advancing toward the goal. This means keeping an open mind about soil practices that can create temporary soil damage, because those practices may ultimately create a more friable soil. Sometimes promoting soil life involves using a disc or tiller to work in composts and amendments, especially when starting a new garden. Other times it includes broadforking a bed to break up compaction, which allows for better water infiltration and soil respiration that in turn promotes photosynthesis—a central goal for growers, as I explain in chapter one. The genius of the broadfork is that, although it causes some significant disturbance in the moment of use, its action can actually enhance soil conditions. And when a broadfork is used in harmony with the guiding principles of caring for living soil, it's a tool that eventually renders itself obsolete.

There are other good reasons to abstain from dogma, too. For one thing, soil science is ever-evolving, and future discoveries could change our understanding of what helps the soil and what hinders it. For another, some practices that *shouldn't* succeed sometimes do, while practices that *should* succeed sometimes don't. One example of this dichotomy is interplanting with carrots, which are not a very competitive crop. Most of the time, sowing carrots around other crops doesn't turn out well for the carrots, and yet, some growers end up with excellent results. Soil biology is profoundly complex and dynamic, and it will take some time to dial in your growing systems and build up your soil's health. At first, you may have to undertake more disturbance than you'd like or more than you see other growers doing. Don't worry about all that—focus on what your soil needs in your context and it will thrive.

Figure 0.2. Living pathways between beds of okra: keeping the soil covered and planted as much as possible.

Make good decisions for your farm business, as well. Run trials. Start small. Test a couple of different methods in a few beds rather than remaking the entire farm with a no-till system you've never tried before. Ultimately, if you're doing things right—keeping the soil planted, covered, and managed with low disturbance—your production and sales will reflect it.

The Original Stewards

Vastly underrepresented both in this book and in conversations about regenerative agriculture are the contributions of indigenous populations—the people who employed the stewardship model of soil management for thousands of years before being dispossessed of their lands or shipped across the ocean and enslaved. Like many Americans, I am descended from colonizers and slave owners. And I firmly believe we owe it to the indigenous and Black populations to avoid claiming their style of agriculture as our invention. No individual alive today is the originator of concepts and practices such as land stewardship, living soil, permaculture, conservation agriculture, or mulching. Being conscious of that can help to repudiate the hubris that led European settlers to violently force indigenous people from their lands and force African slaves to do the work of tending the soil and harvesting the crops. We are simply discovering what indigenous populations knew intuitively for thousands of years: that our role is not to force anything in Nature, but to listen to it, to steward it. In that way, agriculture that focuses on living soil is not an innovation, it's an apologetic response to the many wrongs forced upon the land and for the attendant harm and loss suffered by many people.

At its core, *The Living Soil Handbook* is a book about making that apology to the soil. It's about leaving behind the forceful-agriculture mindset and enabling the soil to do what it naturally wants by once again engaging in regeneration. It's about rebuilding that relationship with the land, studying it, and constantly working to understand it. As in all relationships, you will make mistakes—and as in all relationships, it is recognizing and owning those mistakes that will keep the bond alive.

I'll conclude with this thought: The dusty land deeds and rusty barbed wire fences that define the physical boundaries of farms cannot contain the environmental harms of forceful agriculture. Our waterways are full of eroded soils and leached-out chemicals that originated on farms located miles away. Bird and insect populations are declining all over North America and in many other places around the world. The health of communities is diminishing, and one reason for that is the lack of nutrients in, and the abundance of pesticide residues on, food grown through conventional agriculture—agricultural practices that attempt to force the soil into doing what the farmer wants. The remnants of pharmaceuticals consumed by our sickened communities join the waste stream and, along with nitrates and phosphates from synthetic fertilizer, end up in our lakes and oceans and drinking water.

Chemically farmed soil does not heed borders, but living soil is not fully containable, either. Healthy, vibrant soils clean our water and bring back life. The effects of farms rich with living soil spill out into the communities, too; but instead of sterilizing or poisoning the environment, these farms enliven their surroundings. The populations of birds and bugs that are attracted to a healthy farm environment also enrich the larger ecosystem well beyond the gardens where they reside. Moreover, living soil provides for the grower, economically and emotionally. That's what living soil and no-till are all about: care for the soil and the soil will care for you and your community.

And if you do it right, you'll never grow anything again.

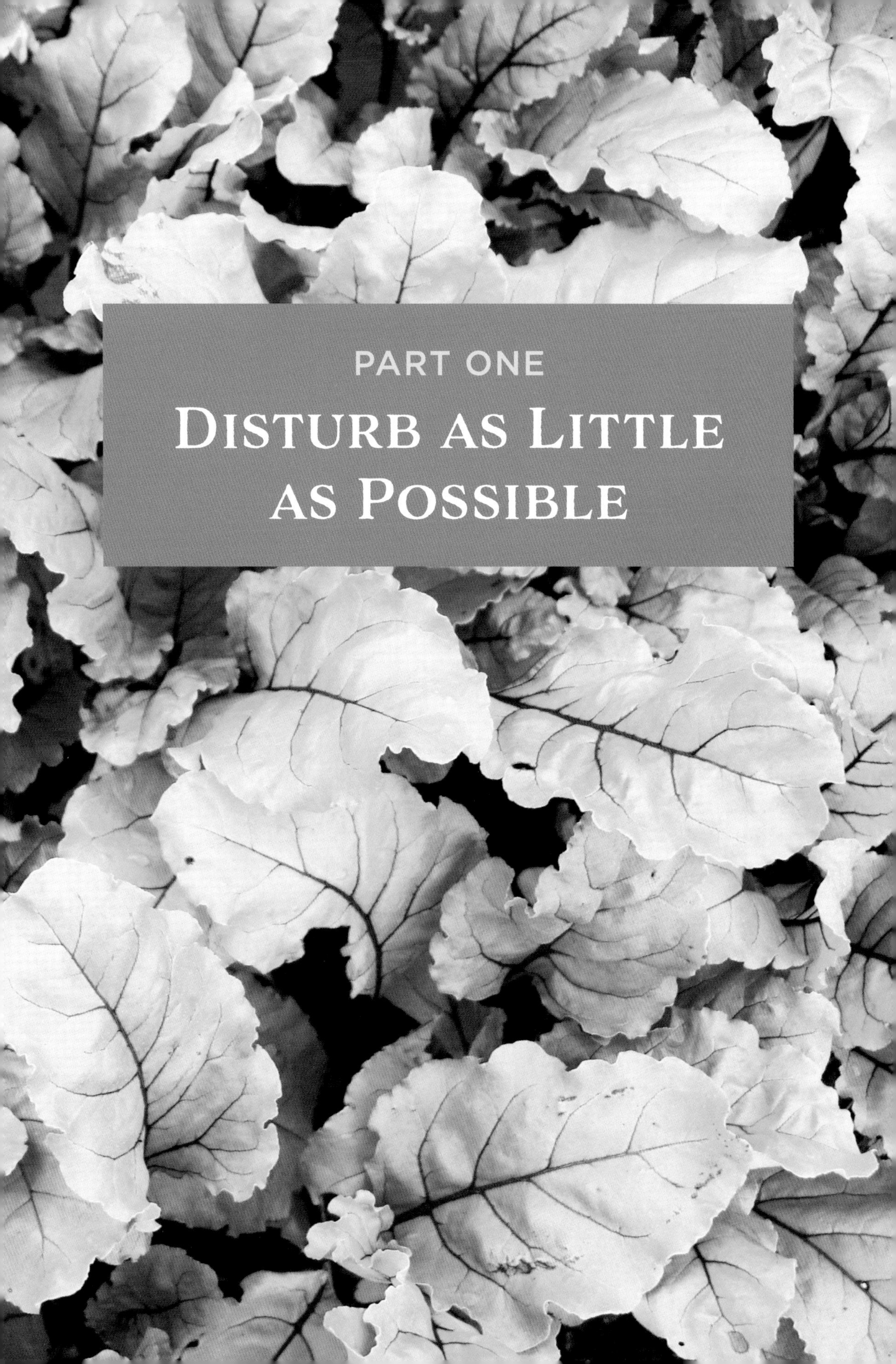

PART ONE

Disturb as Little as Possible

CHAPTER ONE

The Basic Science of Living Soil

The science of what goes on in the living soil is fascinating stuff, but it often gets buried in heaps of jargon that make it unnecessarily confusing. Nonetheless, I believe that every little bit of plant and soil science you're willing to learn can greatly improve your skills as a grower. In particular, diving deeper into the process of photosynthesis and its relationship to the underground ecology will enhance your experience of growing food and can make you a wiser steward of the soil. This is because, trapped inside the scientific gobbledygook surrounding plant and soil science, there exists loads of practical advice on how to manage soil properly. Break through that jargon, and you will find new ways to steward the biology in your soil and grow better and better food. So I'm going to walk you through it. And it all starts about 93 million miles away.

Photosynthesis is quite possibly the single most important chemical reaction on the planet. Full stop. Without photosynthesis, we would have nothing to eat, no air to breathe, no mechanism to cool the atmosphere. There would be no fruit, vegetables, nuts, butter, eggs, or meat. No crude oil, either. Although photosynthesis is a complex process at the molecular level, there are some easy ways to understand how photosynthesis works and why photosynthesis is particularly important to growers—no-till, pro-till, or otherwise. Without question, if you learn one thing from this book, my hope is that it's a basic understanding of the photosynthetic process and its importance not only to plants but also to soil life and health.

Plants are not animals, but they are sitting ducks. In what is arguably one of the most impressive evolutionary outcomes in our planet's history, plants chose to root themselves in place. Indeed, they chose to not run away from potential attackers nor hunt their own food. Instead, plants derive their protection and nutrition in other, more collaborative, ways. And it is

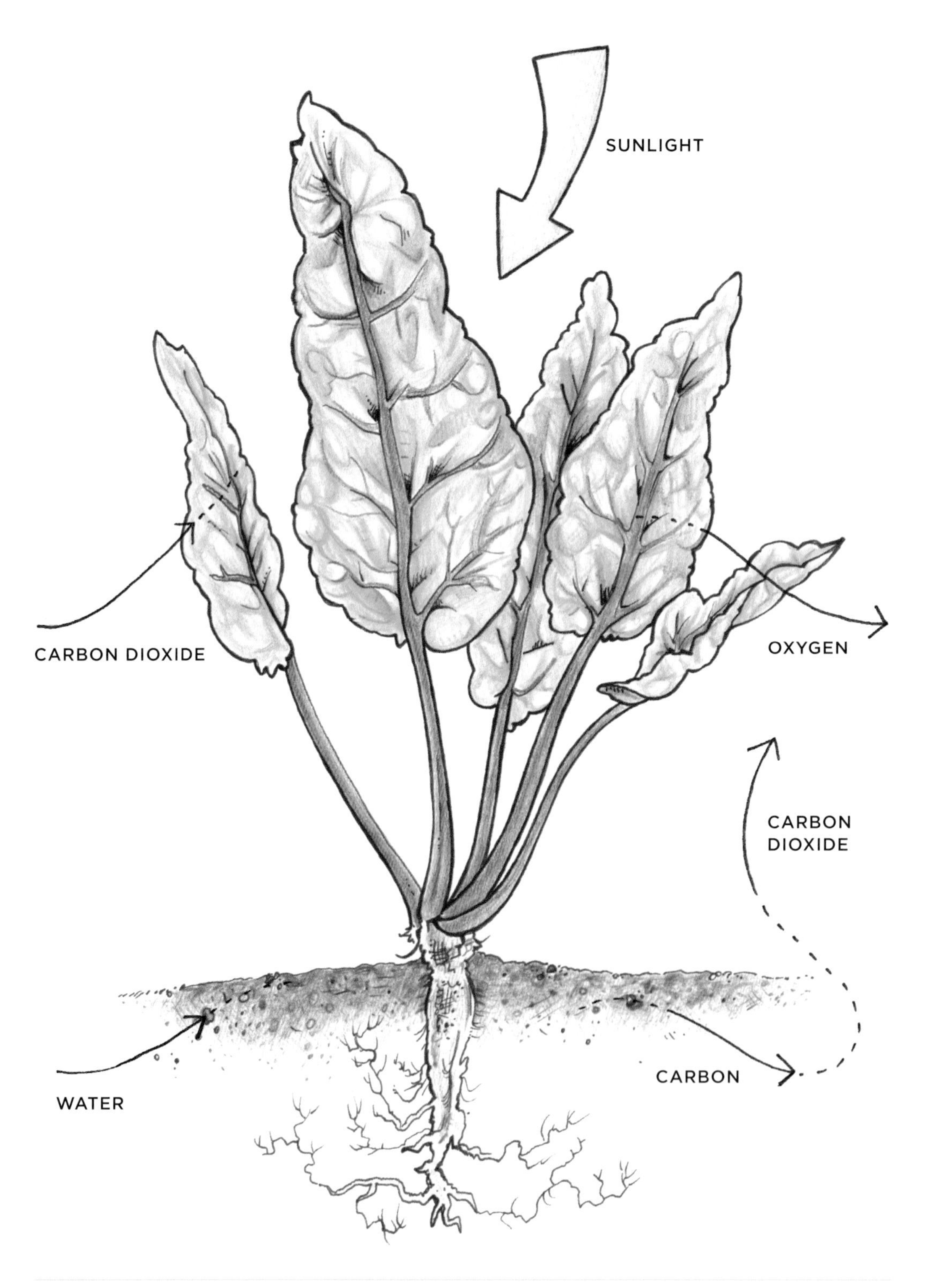

Figure 1.1. Carbon is brought into the soil through the plants but expired back out in the form of carbon dioxide by soil microbes through respiration creating a looping cycle.

photosynthesis that made that choice possible—the synthesis of light into energy. In effect, plants figured out that they could take sunlight, turn it into energy, and trade that energy for protection and nutrients.

In simple terms, photosynthesis is a two-step process involving water, sunlight, and carbon dioxide (CO_2). The first step begins when plants absorb water (H_2O) through their roots. That water is transported into the leaves and specifically into industrious little cells called chloroplasts. Using the sun's energy, the chloroplasts split the water molecules apart into their constituent atoms, which are hydrogen and oxygen. So the plant retains the two hydrogen atoms from the H_2O and releases the one oxygen atom into the atmosphere. Indeed, the air we and other creatures breathe comes from plant cells using sunlight to split water atoms.

Through a complex series of interactions, this initial part of the photosynthetic process ultimately produces two energy-carrying molecules (NADPH and ATP). Next, carbon dioxide is absorbed into the plant through stomata, or tiny pores in the leaves. Inside the chloroplast—the cells where photosynthesis takes place—that carbon dioxide is combined with those aforementioned energy-carrying molecules that were created in the first part of the photosynthetic process. Combining the energy-carrying molecules with carbon dioxide results in various carbohydrates—little molecular bundles of energy—that the plant can use for its own construction. However, the plant doesn't keep all of those bundles of energy for itself. It reserves as much as two-thirds of the carbohydrates it produces to use in a special belowground barter system. Indeed, photosynthesis isn't miraculous simply because it allows plants to use the sun's energy to make food. It's also miraculous because it allows plants to make so much food that they can feed not only themselves but also a complex community of soil organisms. And that is the key to an amazing set of symbiotic relationships between plants and microbes. It is also why photosynthesis is critical for living soil, and why I have dedicated the first section of the book to this subject.

How Photosynthesis Feeds the Soil

Plants can't survive solely on the carbohydrates that they create through photosynthesis. These carbohydrates are sort of the french fries of the plant world—delicious and packed with energy but incomplete nutritionally. Like humans and other complex organisms, plants need a variety of nutrients beyond just carbohydrates for their health and growth. It makes sense that most, if not all, of the 17 essential nutrients that plants need for healthy growth exist in the soil where they've chosen to grow. But plant roots largely lack the ability to harvest these minerals and nutrients on their own. Plants do have the amazing ability to absorb some types of bacteria into their root tips and extract nutrients from them (this is called the rhizophagy cycle).[1] Plants can also utilize some amino acids and other forms of organic nitrogen for this extraction.[2] However, unlike animals, plants can't really "eat" as a way to gain nutrients. Luckily, soil microbes such as bacteria, fungi, and archaea are all equipped with special enzymes that extract nutrients and minerals from soil particles or organic matter. When those microbes die, or are consumed by predators such as amoebas, nematodes, earthworms, or the like, the nutrients that they gathered are left behind in a plant-available form—a form plants can absorb through their roots.

The excess molecular bundles of energy (carbohydrates) that form during photosynthesis are mixed with hormones, organic acids, fatty acids, amino acids, and other compounds, creating extraordinary sugary concoctions called root exudates that plants slowly secrete through their roots. These secretions help nourish the belowground microbial life, and they also can alter the pH and other elements of the environment around the roots. Root exudates can also attract specific microbes that specialize in extracting certain nutrients from the soil.[3] Other exudates can repel the less-desired soil fauna, diseases, and the roots of other plants. Perhaps you have heard that a cereal rye cover crop can inhibit weed growth and seed germination? That effect is called allelopathy, and it comes about from chemicals contained in the exudates of cereal rye and other plants.[4] Generally speaking, however, root exudates feed the soil life and the soil life feeds the plants.

This is what the term *living soil* encompasses—a soil that is filled with microbial diversity and living plant roots exchanging nutrients. This exchange of nutrients is also vital for plant health. Plants can use some of the nutrients they gain in this exchange to create phytochemicals, often referred to as phytonutrients. These complex compounds help plants protect themselves against harmful pathogens as well as oxidation from oxygen or damage from extreme sunlight. When we humans consume plants, we ingest these phytochemicals ourselves and our bodies are able to utilize them in much the same way plants do—as protection against diseases, as antioxidants, and more. Phytonutrients are also largely what give plants their amazing aromas and flavors. Generally speaking, the better a plant is able to access the nutrients it needs and create these compounds, the better it tastes and the more nutritious it will be as a food source. Of course, plants also provide a massive food source for the soil life in the form of carbohydrates. Thus, those microbes have a lot of incentive to defend their plant hosts. Various fungi and bacteria produce novel secondary metabolites that are highly potent chemicals capable of fending off various pathogens (penicillin is arguably the most well-known secondary fungal metabolite).[5]

Predatory microbes, including some fungi, consume harmful microbes.[6] All this adds up to symbiosis at its finest, and it's all driven by photosynthesis and the resulting root exudates.

One other highly consequential phenomenon that takes place among this alchemy and nutrient exchange is carbon sequestration. Carbon sequestration is the act of taking carbon out of the air and securing it in the soil. It is a term that comes up frequently in discussions of climate change as an approach to managing excessive levels of carbon dioxide—a potent greenhouse gas—in the atmosphere. However, carbon sequestration is complicated and sometimes misunderstood. Remember, during photosynthesis plants take carbon dioxide from the atmosphere and mix it with energy-carrying molecules to create exudates. Thus, those exudates are loaded with carbon. That carbon is then injected into the soil through the roots. But the carbon doesn't simply sit there in the soil. In healthy soil, microbes consume and then respire much of the carbon back out of the soil in the form of carbon dioxide. The plants then reabsorb that carbon dioxide again during photosynthesis and start the process over. This is the carbon cycle: In a healthy soil, much of the carbon that enters the soil goes right back out of it. So, is the carbon ever actually sequestered? The answer is a resounding . . . sort of.

Some of the carbon that enters the soil remains there—especially if the soil is not churned up by tillage, as I describe later in this chapter—but the carbon is likely to change forms. For instance, let's say a bacterium eats some carbon-rich exudates seeping from plant roots. The bacterium then uses that energy to multiply itself. Some of the new bacterial cells are later consumed by a larger organism such as a protozoan. That protozoan is itself then eaten by an earthworm. So long as that earthworm isn't plucked out of the ground by a bird or mole, it will live and die in the soil.

During this sequence of eating-and-being-eaten, much of the carbon pumped into the soil by plant roots ended up respired out again in the form of carbon dioxide. But a small part of what began as a carbonaceous snack for a bacterium ended up as part of an earthworm. And when that earthworm dies, some of the carbon in its body will become a more stable form of carbon that persists in the soil through the formation of soil aggregates. Soil aggregates arise when microbes "glue" particles of soil together. In the process, particles of carbon—such as a microscopic slice of an earthworm—can become trapped among those soil aggregates. This encloses the carbon and makes it less accessible for other microbes to consume.

Plant roots are another source of carbon in soil. Roots are highly carbonaceous structures, and when plants are harvested or die, roots left behind in the soil provide food and habitat for microbes for several months, thus

Figure 1.2. This solarized cover crop will eventually add a huge boost of organic matter to the soil beneath.

storing that carbon. All of this activity added together results in accumulations of small amounts of carbon compounds, and we refer to this accumulation as soil organic matter.

Soil organic matter is crucial to soil life and plant health. First, soil organic matter is a major nutrient source. It is consumed by soil organisms, and this process eventually releases the nutrients that were trapped inside the organic matter so that plants can absorb them. For example, a good amount of nitrogen is made available to plants by microbes consuming nitrogen-rich organic matter—nitrogen being a macronutrient that can be difficult for plants to access. With more soil organic matter comes more soil aggregates, which act as microbial habitat. These soil aggregates are full of pores that act like tiny cups to retain water in dry times, but then act like a sophisticated sewer system during downpours, allowing excess water to flow through. (You can think of soil aggregates like a sieve—when you spritz a sieve with water, the water collects between the holes in the sieve, but when you pour water through, the water flows freely.) These aggregates that accompany the organic matter also keep the soil aerated so that carbon dioxide can exit the soil and nitrogen and oxygen can enter. Because it is so important to soil health, organic matter percentage is one of the key soil management metrics. This percentage can be determined by a soil test. I discuss the utility of soil testing and its pros and cons throughout this book, but learning the organic matter percentage of your soil is one clear benefit of having your soil tested. Observing how organic matter percentage changes over time is a simple way to evaluate the efficacy of your soil practices—is that number going up or at least staying stable year after year? If not, or especially if that number is decreasing, the soil may be respiring more carbon than it is accumulating. Under those degraded conditions the soil retains less water, it contains fewer vital nutrients, and it cannot support photosynthesis at a high level.

The Five Keys to Photosynthesis

And so we return to the beginning of this discussion: Without effective photosynthesis, we cannot grow healthy plants and healthy food, nor maintain healthy living soil. Let's examine the practical steps growers can take to maximize the photosynthetic process in the market garden.

Five key factors determine a plant's ability to photosynthesize: sunlight, carbon dioxide, water, soil organisms, and nutrients. The good news is that growers have some amount of control over each one. Though all of these factors work in concert, it's important to keep each one individually in mind as you read the rest of this book.

Certainly all of these factors can be controlled to the nth degree in an automated greenhouse, and to a lesser extent under high tunnels or caterpillar tunnels. Indeed, this ability to control conditions is why more growers are incorporating protected culture into their farming systems—the plants perform well when they have optimum conditions for photosynthesis and are protected from extreme weather events. Even so, I'm not advocating for growing all of your crops in a completely controlled environment such as a greenhouse.

There are many ways to cater to a plant's photosynthetic activity or protect it from weather extremes that don't involve plastic or petroleum products, which may not be as environmentally friendly. One common way to mitigate excessive sunlight, for instance, is to shield plants using shade cloth or high-tunnel plastic. Growers can also strategically plant trees and shrubs to provide afternoon shade for growing areas, which additionally provides the benefits of more photosynthesis. The goal doesn't have to be perfect control of the environment. Instead, focus on making conditions for photosynthetic activity as ideal as possible in your context. Keep that in mind as you read about the five key factors and consider what you can do to best cater to the needs of photosynthesis on your farm.

SUNLIGHT

In order for plants to photosynthesize, they need access to sunlight and they need that sunlight in the proper amounts. When plants receive either excessive or inadequate sunlight, photosynthetic activity slows down or ceases altogether.

An example of insufficient light is when young seedlings are germinated in low light conditions and stretch out in search of sunlight. This elongation—referred to as legginess—results in a weakly structured plant that may not be able to hold itself up. This can lead to problems such as foliar diseases, because drooping stems and leaves that touch the soil can be infected by soilborne pathogens. Of course, too little sunlight can also slow down the growth and production of maturing crops, making them weak, low yielding, and slow to mature. Growers can address low light levels by opening up tree canopies where needed or by providing supplemental lighting in greenhouses.

Plants can protect themselves from excessive sunlight, but there is a limit to their defenses. When exposed to excessive radiation from sunlight, plant tissues (including fruits) can incur physical damage in the form of sunburn, and dehydration and oxidative stress can occur as well. Generally speaking, oxidative stress is not a bad thing, and plants produce antioxidant compounds to combat oxidation. At a certain point, though, oxidative stress can become excessive, causing photosynthesis to cease.[7]

Figure 1.3. Summer beets tend to suffer less stress under the diffused light of high tunnel plastic, which makes them less susceptible to disease.

As mentioned earlier in this chapter, methods to mitigate damage from excessive sunlight include using shade cloth or planting trees and shrubs around gardens. We've found that some tender crops such as lettuce and arugula benefit from shade for a couple of weeks in the summer while they're getting established in the field after transplant. Mobile structures like caterpillar tunnels can be excellent tools for adding sun protection where needed. Beets in particular flourish under the small amount of shade provided by high tunnel plastic in the summer. In several studies, using shade has shown to decrease the percentage of unmarketable (split or sunburned) fruit on crops like peppers and tomatoes.[8] Too much shade is not a good thing, though. A study on purple pak choi found that photosynthesis improved after 5 days of lowered light levels, but dramatically decreased after 10 and 15 days.[9] Interplanting, which is described in chapter eight, is another way to provide sun protection. For example, taller crops such as corn can be used to provide some shade for more sensitive crops like lettuce.

CARBON DIOXIDE

No discussion of carbon dioxide management is complete without mentioning the enzyme called RuBisCo.

Essentially, RuBisCo (ribulose-1,5-bisphosphate carboxylase-oxygenase) is responsible for the first step in fixing carbon dioxide from the atmosphere. But it's not particularly good at its job. Dr. Robert L. Houtz, a horticulturalist from the University of Kentucky who studies this enzyme, told me that RuBisCo fixes carbon dioxide "at a rate that is basically half of what it could be." (Carbon fixation is a general term for the process of converting inorganic carbon into organic compounds; photosynthesis is a prominent example of carbon fixation.) Dr. Houtz as well as many evolutionary biologists believe this inefficiency is largely because RuBisCo evolved at a time when the atmosphere had very high levels of CO_2 and only trace amounts of oxygen. When oxygen levels rose (because of photosynthesis), RuBisCo did not evolve to adapt to the changing atmosphere (though other mechanisms within plants did). So in order to make up for the enzyme's inefficiency, a plant just creates a lot of RuBisCo in its leaves. The only way to increase RuBisCo's rate of carbon fixation, says Dr. Houtz, is to increase the level of carbon dioxide around the plant.

The key to generating high CO_2 levels naturally is fairly simple. In healthy soil that is high in organic matter, soil microbes respire quite a lot of carbon dioxide. This is the natural carbon cycle as described previously. Indeed, showing you how to boost that natural cycle in your gardens is one of the important messages of this book. Recall that plants push carbon into the soil through root exudates, microbes consume that carbon and organic matter, then respire much of it out in the form of CO_2 (not dissimilar to what we do when we eat food). The plant then fixes that CO_2 again and restarts the cycle. Fungi, for their part, are particularly adroit at respiring carbon dioxide, so soils higher in fungi—or garden walkways full of decomposing wood chips—assist in providing CO_2 for RuBisCo to fix. Adding mushroom production to a closed greenhouse may also help replenish CO_2 levels. They say the farmer's footprint is the best fertilizer, and that's sort of true in a greenhouse, too, because humans exhale carbon dioxide—so next time you're pruning tomatoes in a hot greenhouse, remind yourself that

Importing Carbon Dioxide

There are both artificial and natural ways to achieve higher levels of carbon dioxide right around the leaves. As for the artificial ways, some greenhouse producers import carbon dioxide to increase CO_2 levels—this is called CO_2 fertilization or CO_2 enrichment. However, the details for when and how to do that are complicated, and the equipment is often expensive. Furthermore, too much carbon dioxide can have negative effects on plant growth, such as growth reductions and leaf damage. Light levels must also be adjusted accordingly. With more CO_2, plants also require more nitrogen and so soil nutritional status must also be considered.[10] Nonetheless, CO_2 fertilization can be effective. If artificial CO_2 is something you're interested in for your greenhouse crops, first consult with an agronomist.

Other ways to increase carbon dioxide levels include heating a space with stoves such as rocket mass heaters and CO_2 generators. These options generally heat a growing space where desired (conversely, they can be designed to specifically not heat a growing space during summertime). Bringing dry ice into a greenhouse may be a literally cooler alternative, though its efficacy is not well studied.[11]

When introducing extra carbon dioxide in a greenhouse, great care must be taken to protect both the plants and one's self. Excess carbon dioxide levels can be dangerous for humans in closed environments like the greenhouse—this is especially true when using stoves. Extensively research these options for CO_2 enrichment before implementing any of them.

you're also helping to fuel the carbon cycle! To be sure, natural CO_2 sources like these may not compete quantitatively with artificial methods of producing carbon dioxide.

WATER

As I described earlier, one stage of photosynthesis splits water molecules apart, yielding the oxygen we breathe. Water is also the transport system for nutrients and a fundamental part of the cooling mechanism for plants. Too much or too little water can have grave effects on a plant's ability to carry out the rest of the photosynthetic process.

For starters, when water levels are low—such as in a drought or because of inadequate irrigation—specialized cells called guard cells that surround the stomata do not allow those pores to open. You'll remember that stomata are the tiny pores in leaf surfaces that allow carbon dioxide into the plant and water to escape. These "valves" in the leaves help to create a negative pressure, which siphons water up from the roots (if you've ever wondered how plants get water to defy gravity, now you know—the stomata do much of that literal heavy lifting). This movement of water also helps to cool the plant. When water levels are low at the roots, however, the plant closes the stomata to regulate water loss.[12] Generally speaking, the more severe the drought, the more severe the closure. And if stomata are not open, photosynthesis will cease and the plant can overheat.

Alternatively, an oversaturation of water at the roots—caused by flooding or poor drainage—can likewise lead to closure of the stomata and produce other photosynthesis-limiting factors including poor soil respiration, inadequate mineral absorption, and more.[13] This puts a heavy importance on decompacting soils and maintaining proper soil drainage and soil aeration to avoid soggy soils. (For more about soil compaction, see "Starting from Scratch" on page 35.)

Soil-enhancing organisms such as worms also use moisture to help breathe through their skin. Microbes utilize water for transportation, for the enzymatic activity required to extract minerals from rock particles, and more. Floods, stagnant water, or droughts can devastate beneficial soil fauna populations that contribute to plant health.[14] Living soil does not exist without water, so an effective irrigation system coupled with good drainage, high organic matter, and sufficient soil coverage are essential to maximizing photosynthetic activity.

SOIL ORGANISMS

As discussed previously, plants release various root exudates to attract certain microorganisms and manipulate the pH of the rhizosphere. However,

Figure 1.4. As food passes through the digestive system of worms, it is enriched with beneficial organisms and amino acids, leaving that portion of the soil healthier than when the worm found it.

while the plant has some say in the microbes they attract, the microbes also have some say in the plant.

Certain soil microorganisms release specific hormones that can alter the physical structure of a plant and regulate its growth.[15] This means that the microbes assist with designing a plant's architecture, from the roots to the leaves and fruit. Belowground microbes also play a role in aboveground plant defense or susceptibility.[16] Soil organisms can affect the traits of flowers, making them more attractive to pollinators by manipulating the shape and size of the blossoms and also the composition of the nectar.[17]

The important takeaway here for growers is that microbes matter to photosynthetic activity throughout a plant's life—they help keep the plant healthy, productive, pollinated, and in many cases, protected. Regular additions of composts, compost teas, or compost extracts that are rich with beneficial microbes is one way to ensure those populations are retained and grow. For example, we soak every tray of transplants we grow in compost extract or compost tea before transplanting to ensure the presence of diverse microbial populations in the root zone at the time plants are moved into garden beds. (For more on transplanting see chapter eight.)

NUTRIENTS

At the time of this writing, scientists have identified 17 macro- and micronutrients that plants require for survival (see "The 17 Essential Nutrients" on page 22). All of those nutrients play a role in photosynthesis and plant growth. For example, the chlorophyll molecule, which is an essential component of photosynthesis, contains magnesium atoms. So if magnesium is lacking in the soil, plants may not be able to make enough chlorophyll to carry on optimal levels of photosynthesis. Potassium, another essential macronutrient, facilitates the diffusion of CO_2 into chloroplasts.[18] Nitrogen, for its part, plays numerous roles in photosynthesis and is even embedded in our DNA and RNA—no life exists without it. Micronutrients have important roles as well in photosynthesis; manganese, for example, helps the reactions powered by solar energy that result in the splitting of water molecules.[19]

As discussed previously, plant roots absorb the nutrients that microbes harvest. However, it is important that all 17 nutrients are present within the top six or so inches of soil for plants to readily access them, especially at the critical early stages of growth. Some agronomists hypothesize that every nutrient that plants need is already present in every soil. Whether or not you agree with that conceptual framework, the results of a soil test can help to confirm the nutritional status of your soil. Any grave deficiencies in the first six inches of soil can be addressed by adding the missing nutrients

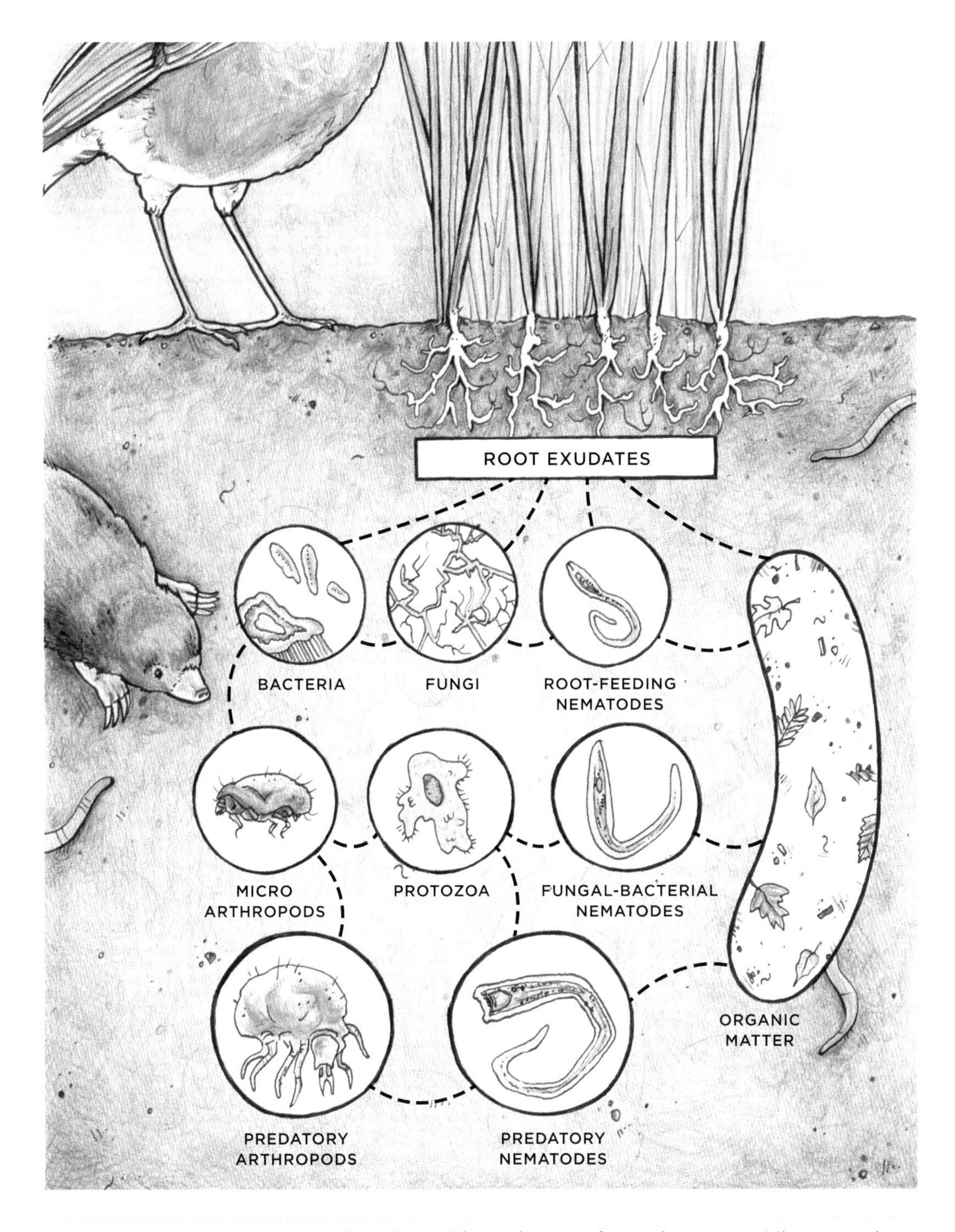

Figure 1.5. The unique food web in the soil branches out from plant roots: Microorganisms consume root exudates, larger organisms such as nematodes and arthropods feed on the microbes, and animals and birds feed on large soil organisms. All of those creatures also produce waste matter, which ends up as a food source for plants and contributes to soil organic matter.

The 17 Essential Nutrients

Macronutrients are those chemical elements that plants need in relatively large amounts. *Micronutrients* are those chemical elements required by plants only in trace amounts.

MACRONUTRIENTS		MICRONUTRIENTS	
Calcium	Oxygen	Boron	Manganese
Carbon	Phosphorus	Chlorine	Molybdenum
Hydrogen	Potassium	Copper	Nickel
Magnesium	Sulfur	Iron	Zinc
Nitrogen			

under the guidance of an agronomist. Cover crops—which are crops grown to enrich the soil—can also help to bring up nutrients that are present in deep layers of the soil. (I cover soil testing, fertilization practices that supply minerals, and cover cropping in more detail in later chapters.)

No matter what you apply to the soil, always be aware that it is soil life that feeds plants, not farmers. Your goal is not to feed the plants directly but to feed the microbial populations in the soil. Synthetic chemical fertilizers formulated as "plant food" can have grave negative effects on soil life, altering or even killing biology in the soil and nearby waterways.

Defining Tillage

Let's consider the impact of tillage on the key factors needed for photosynthesis. A good place to start is to simply define the word *tillage*.

Many dictionaries describe tillage as "preparing the ground to grow crops." For centuries, that's all it was. Dating back to the earliest agrarians, indigenous farmers prepared small plots of ground by hand or with animals, using implements made of stone, bone, wood, and later, metals. Certainly, many of these traditional practices opened up and exposed the soil temporarily. The scale of farms was generally small. Fields were regularly fallowed—that is, they were encouraged to go back to hosting wild plants—several years at a time, which replenished nutrients and repaired soil structure. These practices served farmers for thousands of years, and many indigenous cultures still practice these small-scale growing methods.

Over time, though, the scale of farming changed, and with it, the definition of tillage shifted. The development of new tools enabled farmers to open up and plant larger and larger plots of land. Cast-iron implements replaced wood implements, and then steel replaced cast iron. Powerful tractors entered the picture, followed closely by chemical fertilizers. Changing practices in farming led to huge swaths of exposed soil, and an increasing potential for soil degradation. Unfortunately, that potential has been realized many times over.

The Dust Bowl that hit the Great Plains is one tragic example of the consequences of soil degradation. After dispossessing Native Americans of millions of acres of grassland, European settlers and their descendants spent several decades plowing up the native prairies right before a long drought hit the region in the 1930s. The drought lasted almost the entire decade. (NASA recently determined that 1934 was the worst drought year in the last millennium.[20]) Without the strong roots of native prairie grasses to hold the soil in place, the drought caused colossal dust storms that dropped prairie soil as far east as Washington, DC and New York City.[21] Although the Dust Bowl did result in some government action, such as the creation of the Soil Conservation Act and the Soil Conservation Service (later changed to the Natural Resources Conservation Service or NRCS), dust storms and desertification caused by agricultural practices have not entirely disappeared in the United States. At least one recent study demonstrated that dust levels are once again rising to dangerous levels in the western United States, largely caused by agriculture.[22]

Without a mulch covering or living plant roots to hold the soil in place, soil becomes susceptible to erosion from wind and rain. Intense heat and sunlight can also dry out soils and burn up organic matter. A 2020 study examined soil loss in 38 countries and found that more than 90 percent of the conventionally farmed soils studied were becoming thinner (losing topsoil), and 16 percent of them had lifespans of fewer than 100 years left if the methods used to tend those soils were not changed.[23] Some media reports, as well as leaders from the Food and Agriculture Organization (FAO) of the United Nations have suggested that many agricultural soils have as few as 60 years left.[24] All that to say, the devastation wrought by tillage and conventional agriculture is extensive, ongoing, and ultimately leads to less farmable land.

To understand how large-scale mechanical tillage begets soil loss, think of soil as a major underground city. Like all cities, soil requires infrastructure. It needs tunnels for the transport of air and nutrients, and it needs housing for its residents (soil organisms). It needs a stable physical structure that allows water to flow laterally and to drain vertically. In living soil, plant roots and soil aggregates bind the soil together, creating vital stability. Earthworms carve tunnels, making it easier for air to come in, carbon dioxide to leave, and fungal hyphae and plant roots to thread their way through. However, when we crush that infrastructure by tilling and poison that soil life by applying pesticides or chemical fertilizers, we render the soil vulnerable to erosion.

Soil organic matter is plundered in major tillage events. Soil aggregates are broken apart and oxygen is simultaneously whipped into the rhizosphere. Newly enlivened oxygen-loving bacteria begin to feast on soil organic matter and respire it as carbon dioxide. Because there are few or no plants present to capture much of that carbon dioxide gas and return it to the carbon cycle, it flows into the atmosphere unobstructed.

Mechanical tillage is catastrophic to fungal populations. Bacteria are highly adaptable to changing conditions, but many fungi require significant time and energy to build a mycelial network. Mechanically churning the soil rips all of that apart and the fungi must begin again, starting from the level of individual spores. Nematodes, arthropods, earthworms, and other predatory organisms likewise get pummeled when organic matter is lost by being tossed onto the soil surface where it burns up in the sun, blows away, or is swept off by heavy rain. The end result is a soil rich with bacteria, low in predators, and with damaged fungal populations. This type of soil primarily favors the growth of plants that a soil ecologist might praise as adaptable, but that farmers would condemn as unwanted weeds.

Not only does tillage create conditions that some weedy species find ideal, it also brings weed seeds to the surface of the soil, which encourages them to germinate. Remember, weeds are simply plants that farmers aren't intending to grow. Some weeds are edible, some have excellent soil-enhancing properties, and many of them are harmless. Weeds can be fast growing, and they are often prolific producers of seed heads or underground rhizomes. These qualities are not a negative in terms of restoring damaged

Mycorrhizal Fungi

Mycorrhizal fungi latch on to or inhabit the roots of many kinds of plants and trade moisture and nutrients with those plants in exchange for root exudates. These fungi also send rootlike structures called hyphae through the soil, and these hyphae act like root extensions for the plants. These fungi can mine the soil for nutrients and water 100 times deeper and farther than a root can while also helping to fend off soilborne diseases. In forests, these fungi-plant associations, which are known as mycorrhizae, can act much like an underground internet, sending nutrients and messages across large distances among plants. Keeping these networks intact so they continue to grow, spread, and collect nutrients is an enormous part of the incentive to not overly disturb the soil.

Figure 1.6. Living soil is key to growing great food, and fungi, here growing alongside a row of beets, are one of the few reliable indicators of biologically active soil

soil. Weeds come to the rescue when soils are barren. They may be indicative of what our soils lack in terms of nutrients or physical properties.[25] Effectively, Nature relies on weeds to keep the soil covered with photosynthesizing plants where the farmer fails to do so. Of course, some weeds are aggressively invasive and can compete with crops for moisture and sunlight, which in turn can slow crop growth and result in poor yields.

The ecological risks that accompany regular plowing and mechanical tillage don't stop at microbial devastation, organic matter loss, and erosion. If the soil is worked when it's too wet, tillers, plows, and other similar implements can create various types of compaction in the soil. These types of compaction are a significant barrier to creating healthy soil, and they limit crop production and soil health tremendously. Surface compaction occurs when bare soil is exposed to heavy rains or foot traffic. This form of compaction inhibits water penetration and limits respiration.[26] Hardpans are compacted layers that generally form at the greatest depth a farm implement reaches, and these deep hardpans can inhibit root and water penetration. This type of compaction is stubborn and can persist for many years even after a farmer adopts good soil management practices.

Many bacteria and some fungi can survive in compacted soils, but microbial predators such as nematodes often suffer. Without those predators, bacterial populations increase but nutrient availability may not.

Moreover, if water cannot properly drain, photosynthesis slows or stops entirely while microbes and plant roots drown. The microbes that survive such saturated environments begin to consume the available nitrogen, leading to nitrogen loss (denitrification). Both carbon dioxide and hydrogen sulfide gasses may build up and become toxic to plant roots.

With these negative impacts of tillage in mind, let's explore the idea of expanding our concept of tillage. Tillage is not solely the outcome of using a tiller or disc or plow. Many other kinds of tools can cause some or all of the problems we generally attribute to mechanical tillage. It is not the tool but the user who determines whether a particular act of tillage creates minor soil disturbance or major soil disturbance. Put another way, it is not the tool that decides how to till a piece of land, how long and how intensively, and how to follow up after tilling—those are the decisions that farmers make, and that determine long-term soil health and performance.

To underline this point, let's compare and contrast some tillage tools and how they can be harmful or helpful to the soil depending on how they're used. First on the list: the rotary tiller. When used as the primary means of soil preparation year after year, this tool can absolutely devastate soil ecology and structure, repeatedly encouraging all the aforementioned issues. That said, in some cases, a tiller may be ideal for starting a garden. When a farmer or gardener uses a tiller appropriately, this tool can break up existing compaction layers and inject composts and amendments into depleted soils, rapidly preparing them for production, and thus improve that soil's potential to support photosynthesis.

Consider another—sometimes controversial—tool in the no-till world: the broadfork. This large fork has long tines and farmers use it to gently decompact soil. The farmer stands on the crossbar to force the tines into the soil, then steps off and pulls back on the handles, lightly lifting the tines enough to simply crack the soil surface. However, the broadfork can easily be used to heave up large chunks of soil and flip them over, breaking the soil apart and damaging fungal populations, exposing carbon stores, and injecting large amounts of oxygen into the soil thus encouraging an organic matter feast by bacteria.

A less obvious tillage tool is the silage tarp, which is popular for covering an area of prepared soil to cause weed seeds to germinate and then die, or over a crimped cover crop to help terminate the cover crop. But when left in place too long, a silage tarp can also be a form of tillage. Tarps are heavy and can create surface compaction. Exposure to UV radiation causes the tarp fabric to break down and shed microplastic particles, which can be harmful to soil life. Polyethylene tarps do not allow for much gas exchange, potentially creating an anaerobic environment that may encourage pathogenic microbes. This practice may not look like classic tillage—it does not invert or blend soil layers—but it has some of the same negative effects that we associate with tillage systems.

All that to say, when we're working toward a no-till, living soil system we need a clear definition of what we're trying to avoid. We also need a more comprehensive representation of what constitutes tillage. In the English language, words can flip meanings from one generation to the next. *Awesome* used to mean something that evoked terror. *Egregious* used to mean distinguished. The meaning of tillage has flipped, as well. Tillage no longer simply means preparing ground for growing crops. Increasingly, tillage is a set of practices that make the soil less capable of growing anything at all. I suggest that we call tillage what it is: anything done to the soil that does not ultimately promote soil health.

I want to emphasize, however, that no-till growing is not about avoiding *all* disturbance of the soil. Every organism in the soil causes some amount of soil disturbance, even if only at the molecular level. If you were to watch a time-lapse video taken of the soil over the course of several years, the soil would not simply be sitting still. No, the soil would look like it was churning. That churn is what biology is doing at all times in healthy soil. Therefore we need to distinguish the disturbances that benefit the soil from those that don't. No-till represents the former—it is about promoting appropriate soil disturbance and biological stewardship. We do not need to think of ourselves exclusively as destructive organisms. Just like plants or earthworms, we humans are capable of enriching our environment if we so choose. We may not always agree as growers on what that looks like, which is fine. Each soil may require a unique approach that involves differing amounts of initial or sustained disturbance. Some farms may not ever need to broadfork. Others may require a few consecutive seasons in which every bed is broadforked. In some circumstances, bringing in a subsoiler—an implement that rips deep into the soil without turning it over—is the best way to remedy a hardpan layer. In effect, to achieve living soil, we have to be guided by our own observations of what our individual soils need, not by what is popularly believed to be the ideal way to manage them.

PRACTICAL REASONS NOT TO TILL

Along with the abundance of ecological reasons not to till, there are myriad practical reasons, as well. Listing them all might be a book in itself, so I will focus on a few of the ways that changing to a no-till system altered our work and our lives for the better.

To many growers, a no-till system may seem like it must be *more* work than tilling. In some instances, they're right, especially during the first few seasons of establishing a new garden. But there are also many ways in which the types of no-till agriculture I describe in this book can ease the toil of farm work for the farmer and farmworkers. In a successful no-till system, the farmer takes care of the soil, and the soil also takes care of the farmer. Let's see how that looks in practice.

Fewer Weeds

It's a cliché, perhaps, but Nature seemingly abhors bare soil and fills it with living plants. In no-till agriculture, we follow Nature's lead. Weed pressure is dramatically reduced in a no-till system when we stop bringing weed seeds to the surface through mechanical tillage. Simply by keeping the soil covered with mulches and plants at all times, we avoid stimulating Nature's impulse to grow weeds.

As weeds disappear, so does the labor involved in managing them. Labor is the biggest expense on most farms (yes, even if you're the only employee). Before our farm went fully no-till, cultivation took up full days every single week of the growing season. The task became expensive and exhausting in the Kentucky heat. Now that we rely more on mulches to do weed control for us, our gardens require little more than a light spot-weeding every 10 days or so, mostly in the pathways. That's an enormous difference for our farm in terms of time saved, and time is incredibly valuable to us as a family.

Figure 1.7. Here are two views of one of our fields: bare dirt at planting time when we were tillage growers versus our no-till practices, with the soil covered by various kinds of mulch.

There are no- and low-till systems that do require consistent cultivation, especially systems that do not rely on mulch. Weeds are an inherent part of market gardening regardless of what system you use—weed seeds blow in on the wind, are dropped by birds, or fall from boot treads into garden pathways. However, any reduction in soil turnover while keeping the soil mulched as much as possible will decrease weed outbreaks. Moreover, weed seed predation by birds, rodents, and insects is increased in organic systems because wild creatures are less likely to be poisoned or lose their habitats due to tillage or pesticide use.[27]

Longer Growing Season

One helpful benefit of not working the soil—especially in a deep-mulch system—is the ability to plant earlier in the season or when the soil is wet. Indeed, our farm schedule, not the weather, is the greatest determining factor for when we sow new crops. When we were tillage farmers we prayed a lot for a dry day in springtime so we could work the soil. Now, however, weather is no longer a significant determinant of when we can plant or sow.

Of course, we are still limited by moisture and warmth to an extent. For instance, after a heavy rain we have to wait a day or two to broadfork a bed, because the broadfork will simply slide through the soaked bed and lose its effectiveness. As well, we have to wait for the soil to warm up before we transplant summer crops like tomatoes, because its roots are very sensitive to soil temperatures. But because of the way we mulch our beds and pathways, we rarely come into contact with mud, and our feet do not sink into the soil and thus compact it. The dark compost mulch we use warms the soil slightly, as compared to soil that's left bare or is blanketed with a light-colored mulch. This way of farming makes springtime significantly less stressful, too, because it eliminates the need for us to cram in all of our planting during brief periods of dry weather. Now, we can do planting tasks nearly every day—rain or shine.

Greater Focus on Planting and Harvesting

In an ideal farming system, planting and harvesting should always be the bulk of the work, and this is one area where the practical side of no-till really shines. Without weed pressure to worry about, after we plant a crop we can largely leave it on its own until harvesttime, and when harvest begins, we're not fighting through weeds to pick produce.

Going no-till has also dramatically affected our replanting process. In our previous tillage system, crop termination involved some version of mowing, tilling, and then waiting a couple of weeks for the previous crop residue to break down before we could harrow the bed and replant it. To be clear, in those days we couldn't have just skipped the tilling. It was a necessary step to reclaim the bed from weeds and inject amendments, and also to break up surface compaction that had formed on bare soil (we also simply didn't know any other way to farm). During the weeks of bed recovery after tilling, the bed wasn't producing a crop, which reduced our earning potential. Nowadays, we transition from one crop to the next within a matter of *hours* because there is no need to till up weeds and plants or break up surface compaction. We simply cut out the previous crop, spread compost if necessary, and replant.

This change in bed turnover and general management also means we spend less time enduring the noise and physical stress of using our BCS walk-behind tractor. This is another notable bonus of no-till. We still use the walk-behind tractor to mow beds with a flail mower attachment, but we don't have to switch out implements multiple times—mowing, then tilling, then power harrowing—before planting.

Time Savings

When you are not spending your days fighting weeds, or waiting for the soil to dry out, or switching implements, or making multiple passes with the tiller, you suddenly have more time to dedicate to other things. You might decide to improve areas of your farm or work on projects you've been neglecting. Or you might simply enjoy more time in the company of family and friends. This latter point should not be brushed aside. Because farming often takes place in rural, less-populated areas, it is inherently isolating. Farming doesn't always allow for weekends off as other jobs do. Many times, farmers miss out on summer parties and weddings and concerts, and that can be extremely hard on the farmer. I often think about the results of a massive ongoing Harvard University study on life satisfaction. This study started with 268 Harvard sophomores in 1938 and has tracked those students and their offspring, as well as control groups from lower-income urban residential areas, for the last 80-plus years. The data revealed that, more than money or fame, what brought people happiness throughout their lives was close relationships. According to a description of the study results in the *Harvard Gazette*, "Those ties protect people from life's discontents, help to delay mental and physical decline, and are better predictors of long and happy lives than social class, IQ, or even genes."[28] Healthy food is important to long-term health, and that's certainly why many people start growing their own food to begin with. But if your livelihood of growing healthy food denies you the opportunity to build and maintain good relationships, then you're negating some of those benefits. If this describes you, you need to find a growing system that allows Nature do more of the work.

NATURE AS THE MODEL

When learning how to best take care of your soil, you should couple every book or article you read or lecture you attend with a simple walk outside to look around and observe. Let the knowledge you absorb from experts or your fellow farmers resonate with what Nature has to teach you.

Cliché as it may sound, the most successful template for robust photosynthetic efficacy can always be found in the woods, in a meadow, in a prairie, in Nature. For hundreds of millions of years, the microbes we rely on to grow our food evolved under some basic conditions that are still Nature's model today. We can see these conditions in the wild and, in many ways, we must work to replicate them in the market garden.

A few common characteristics are visibly evident in most high-functioning soil ecosystems. First, in forests and grasslands alike the ground is covered and cooled by a diverse plant canopy, often in multiple layers (that is, trees, understory, ground cover, mosses, and so on). A mixture of plant species photosynthesize and feed the soil, cool it, and provide habitat for insects and wildlife. For healthy soil, the quality of the plant canopy is a critical element.

In a natural, plant-based system there is also always some form of organic mulch or thatch covering the soil. This material may be grass stems smashed down by grazing animals on an open plain or leaf mold collecting

Figure 1.8. Fungal populations thriving in our no-till system.

Figure 1.9. We like to provide habitat around the gardens for pest-eating birds such as sparrows, bluebirds, barn swallows, and purple martins.

on the forest floor. Either way, systems that rely on plants almost always possess some sort of vegetative armor that covers the soil. This material helps retain moisture, reduces compaction from impact of raindrops and animal hooves, protects against erosion, and provides fodder and habitat for soil biology, among other benefits.

Speaking of habitat, the presence of wildlife in any ecosystem, including a farm ecosystem, can help ensure that tender young seedlings and transplants succeed. The transition from seed to plant can be a hazardous time for young seedlings. Wildlife such as birds and snakes and amphibians and beneficial insects all help to reduce populations of pests that might pick on tender young plants, and they also combat macro-level garden pests such as mice and voles. Even in the best soils, healthy transplants can require a stress-free establishment period after planting, so it is important to have wildlife around that can swoop in for support.

For us as growers, achieving good photosynthetic activity comes down to replicating Nature's system as much as possible—diverse plant canopies, often with perennials mixed in, complemented by a mulch of some form. Such an environment naturally encourages habitat and fodder for predators, which is a good thing. It's Nature's model, and we have a lot of options for how to emulate it in the market garden . . . right from the start.

CHAPTER TWO

Breaking New Ground

Context is everything in farming. One person's approach to starting gardens in, say, sandy soil may not be what makes sense for you if you farm in red clay, and vice versa. In this chapter, I discuss how to start gardens from scratch as well as how to transition old gardens to a no-till system. How you start your garden will determine much about how it performs for many years to come. In some cases, a zero-till approach to the process is fine, even optimal. In other cases—I would venture to say *most cases*—some amount of opening up the soil will be necessary. A garden needs to be built on a sound foundation just as a house does. When the foundation is weak, eventually you will spend a lot of time and energy fixing it. Plus it's a lot harder to access the foundation after you've built a house, or in this case a garden, on top of it. If the native soil was not prepared well, you may struggle after the fact to correct problems with poor water movement, root penetration, and nutrient availability, all of which can reduce photosynthetic activity, which then leads to a reinforcing cycle of declining soil health.

One conundrum is that you may not have figured out yet exactly what your no-till system will be. Will you work with deep compost mulch? Straw? No mulch? No idea? That is a perfectly fine place to be. Throughout this book there are tips for how to figure out the details of a system based on your context. Regardless of the no-till methods you ultimately choose, you will first need to address the condition of the native soil. So let's start there.

Site Selection Considerations

For some growers, the nature of the available land will determine where the gardens go. Maybe you have only your backyard, or only a single flat area, or perhaps none of the available land is cleared or flat. You can work with it anyway. Throughout history, people have started gardens on steep hillsides, on rooftops, even "floating" on lakes in the case of ingeniously constructed gardens, called chinampas, in parts of Mexico. Where there's a will, there's a garden. However, if you have any amount of choice over where to situate your gardens, there are some important points for you to consider.

First, choosing a mediocre garden plot closer to your house is more optimal than a perfect plot a mile away. Distance spent traveling to and from a garden adds up. Four trips a day at even 2 minutes a piece is a lot of wasted time. If you spent that amount of time just traveling for 300 days in a year, it's 40 hours—hours you could be doing something productive or fun. Distant gardens are also likely to suffer more deer damage and possibly other problems, too, because a farmer is *less likely* to visit a garden that is far away from the house.

In choosing a site:

- Look for full sun exposure year round. Avoid sites where large buildings or trees on the south side could shade the garden in spring and fall, shortening the growing season. The presence of small trees can be an asset when growing some crops, both for shading and as a windbreak, but you'll need to think that through strategically. Because every region has different perennials that work best, I suggest chatting with local farmers and local nurseries about what might work best, and what species of plants have root systems that won't be too invasive.
- Avoid sites in floodplains. All of your work could be lost in a single large rain event.
- Watch for poor drainage. Visit your potential site the day after a large rain and identify any areas that fail to drain. Poor drainage can be improved with soil practices to some extent, but consistent and substantial water accumulation may require the installation of a physical drainage system to move water away from growing areas.
- Consider water access for irrigation. Does the site have a well, a spring, a pond, or county water that can be tapped for irrigation and washing? How much will it cost you to get it there versus other locations? Are there legal obstacles to your use of the water (water rights issues)? Do not assume.
- Evaluate the road access. Access is important for bringing in tools and transplants, and for bringing out the harvest, as well as for receiving large truckloads of compost or wood chips, not to mention hosting customers.
- Consider the slope. Many growers don't have access to flat land (it's basically nonexistent in my area of Kentucky). Take into account any leveling, terracing, or contour creation you'll need to do before you start growing. Slopes of roughly 10 percent or greater should be

modified or terraced in some way to increase water and soil retention. To figure out percentage slope, divide the rise (elevation change) by the run (horizontal distance along the slope) and multiply by 100.

- Identify noxious weeds—bindweed, bermudagrass, and others can present major obstacles to food production. If they can be avoided, do so.
- Determine soil type. Soil type should not be a deal breaker. (See chapter seven for an explanation of soil types.) If you have access to nice loamy soil, great. Many farmers don't. Any type of soil can be improved, and I discuss how to improve soil later in this chapter.
- Test for lead and other contaminants. Lead is readily bioaccumulated by plants and can create toxicity in leaves and fruit. Because lead was a substantial ingredient in paint and gasoline for many years, it has found its way into many soils. Urban areas with dense housing and traffic tend to have soils with higher levels of lead than soils in rural areas, but lead contamination is possible on any site where buildings may have existed.

Starting from Scratch

From weed pressure to disease, it's possible to preemptively manage many issues right from the beginning of a garden's life. It's simply a matter of taking the proper steps . . . and eschewing unnecessary dogmas.

One dogma I caution against is "never till." As mentioned previously, for some plots, it may be entirely possible to start a garden without having to open up the soil at all. In other situations, however, a one-time tillage may be precisely what your garden and crops need for success.

STEP ONE: THE SOIL TEST

A proper soil test is a simple but important step in deciding how to start a garden (and testing your soil regularly over time is a good way to monitor progress toward creating a living soil). The results of a soil test will reveal its organic matter content, any severe nutrient deficiencies, its pH level, the soil's ability to retain nutrients (this ability is termed cation exchange capacity), and more.

I recommend spending the money to work with an agricultural lab that uses the Albrecht method of soil testing, also sometimes referred to as the Base Cation Saturation Ratio (BCSR) system. I don't want to get too deep into the weeds of soil chemistry here. Simply put, the Albrecht method is a type of soil test that assesses the available nutrients in a soil as well as its organic matter (OM) content and pH level (among other soil attributes). The labs that offer this testing often also provide recommendations for organic

Figure 2.1. Soil test results are valuable for assessing the soil at a new site, but remember to visually evaluate your soil too. The presence of earthworms is an important indicator.

macro- and micronutrient amendments (as opposed to synthetic) to "balance" a soil's physical properties. This might be done through such methods as adding peat moss to improve tilth and increase moisture retention.

There is some debate about this approach to soil management. As I stated in chapter one, although some agronomists contend that soil balancing (also called mineral balancing) is an effective means of guaranteeing that plants will have immediate access to the nutrients they need, not all agronomists endorse balancing soil by applying rock minerals and other nutrient amendments. Moreover, the academic research available for BCSR is not entirely supportive of soil balancing (when yield is the measurement). Some agronomists suggest focusing more on incorporating beneficial biology than mineral balancing.

The reason I recommend paying the higher fee required for a BCSR-type analysis is because it is far more comprehensive than the tests available from your local extension service, and the recommendations you receive will be for nonsynthetic amendments. When you receive your test results, check to see whether there are any major deficiencies in macro- or micronutrients. If a nutrient doesn't show up in the test, or it shows up in extremely low amounts, then you know it will be difficult if not impossible for plants to access them. Soil test results also reveal whether any nutrients are present at excessive levels, which will give you an idea of what not to add. In all cases, to avoid creating any other imbalances or toxicities, consult with an agronomist before adding amendments to your soil.

The lab you choose will provide instructions on how to take an accurate soil sample and how to package it up to send in for analysis. For example, some labs accept only dry soil samples. Spring and fall are generally the most desired times to collect soil samples. Avoid sampling in a drought.

Define the plots you want to have tested. Each plot should be no larger than half an acre—the smaller the better. For each plot, you will collect several small samples. Buy or borrow a soil probe to collect soil samples (see figure 7.2 on page 134). Shovels work, too, but a probe provides a more accurate sample. Plunge the probe into the soil about 10 inches. Give the probe a full twist while it's in the soil then pull it up. Remove and discard the bottom inch or so of the core and also the top two inches. Then reserve the remainder of the sample. This is a sample of the rhizosphere—the area around the roots—which is the most important segment for the purpose of growing annual crops. The condition of this portion of the soil indicates how your crops will perform. Mix all the samples from a plot together in a nonmetal container (such as a clean plastic bucket). Make notes about how you selected your samples so that for future testing you can follow the same protocol as closely as possible. After you have assembled the samples from all the plots you want tested, package them up and ship them per the lab's instructions.

I recommend getting an analysis before building your garden beds, because soil test results provide an excellent snapshot of the state of your soil. Tests that reveal your soils to be severely depleted of organic matter, or show some significant nutrient deficiencies, or reveal an extremely acidic or alkaline pH level, present problems that are easiest to remedy before you build your beds. Try to talk by phone with an agronomist from the lab, even if that consultation costs a little extra. Most of these labs do not sell products, and their staff should be able to provide you with an objective opinion. Continue to have samples analyzed over time. These tests will show you how well your soil management efforts are working to improve or maintain soil organic matter. Always work with the same lab because each lab has their own approach to analysis.

STEP TWO: CHECKING COMPACTION

Few physical characteristics of the soil can negatively affect plant growth more than compaction. As described in chapter one, compaction is a hard layer of the soil that is difficult for roots to penetrate. Compaction can lead to poor soil water movement and a proliferation of disease organisms in the soil. Compaction is most common in heavier soils, ground that has been cultivated when overly moist, or ground that has had continuous concentrations of livestock or machinery on it. Sandy soils do not compact as easily as clay soils. Heavy rains on bare soil also cause compaction, as can standing water,

FOCUS

No-Till Beds at Rough Draft Farmstead

Hannah and I farm three-quarters of an acre of diversified vegetables in central Kentucky. Over the years we have tried every no-till method described in this book, and we trial new methods every season. However, we are a production farm, and there are a couple of soil management methods we rely on for guaranteed harvests.

The primary method we use is what I refer to as shallow compost mulching. This is done by simply maintaining a four-inch layer of compost on bed surfaces all season as a thick mulch. Many growers suggest using much deeper layers, but in our experience the compost available to us does not hold

Figure 2.2. Summer tomatoes growing in a terminated crop of rye, crimson clover, and hairy vetch.

among other elements. Most previously cultivated farmland is likely suffering from some amount of soil compaction, and soil that is overly compacted can greatly inhibit crop production for many years. Take this issue seriously.

The most accurate way to measure soil compaction is with a penetrometer. This is an instrument that measures the pounds per square inch (psi) of pressure required to penetrate the soil. Any soil measured at or above 300 psi is considered to be compacted enough to greatly inhibit plant growth. Borrowing a penetrometer is cheaper than buying one. To save money, this tool would be a good candidate for purchase collaboratively with several other farmers.

moisture well enough to apply a layer that is much deeper than four inches—the plants growing in it would be water-stressed. With the thinner layer, our plants have more access to the native soil. And thanks in part to the compost mulch, the native soil holds moisture extremely well.

The other method we increasingly utilize is a cover crop mulch, also known as grown-in-place mulch. Beyond improving soil health, specific cover crops produce long lasting mulches that can block weeds and help retain soil moisture. For this method, we sow the desired cover crop into a shallow compost mulch bed, terminate it by mowing it or covering it with a tarp, and plant as soon as the cover crop has fully died.

Figure 2.3. In most beds, we plant three or four crops in succession over the course of the growing season.

Alternatively, one can use a section of rebar or a shovel to measure soil compaction (I prefer a spade). Go to an area you know to be relatively compacted due to heavy foot or animal traffic and press the instrument into the soil. That will give you an idea of what compaction feels like when using that particular instrument. Then move to the site where your garden beds are or where you want to make beds. Insert the tool or instrument there and compare how far down it sinks without having to apply as much force as you did for the compacted soil. You may immediately be able to tell that the soil at your chosen site is compacted—or that it is not compacted at all. Sometimes an instrument easily penetrates several inches deep, but then stops suddenly and completely. This indicates a compaction layer well below the surface, and the problem will need to be addressed at that level.

Test multiple areas in a plot for compaction. If there is a consistent trend of compaction, you'll need to combat that compaction before you make beds. This is a situation where a never-till attitude will not serve you well.

STEP THREE: AMENDING

With soil tests in hand and your probing for compaction completed, the next step is to figure out what amendments to add and how to apply them. You can follow the amendment recommendations at the rates suggested by the lab. Or you may decide to pursue the task by using biological amendments and biostimulants (in chapter seven, I discuss the use of amendments and creating a fertility program). If you plan on certifying as an organic farm—even if not now but sometime in the future—make sure that everything you decide to add to your soil is listed by your certifier (or by the National Organic Program (NOP) in the United States) and keep records.

Spread amendments evenly over the soil surface. If your soil organic matter is low (below 3 percent or so), consider adding compost, too, and working both the amendments and the compost into the soil with some form of one-time tillage as described in the following sections of this chapter. Before doing any mechanical tillage, consider adding activated biochar—a stable, charcoal-like carbon material that serves as an excellent microbial habitat. Inoculate the biochar by soaking it in compost tea or mixing with a compost for several weeks before turning it into the soil. (Compost is discussed in chapter three.)

STEP FOUR: MANAGING COMPACTION

Hannah and I did not investigate soil compaction well enough on our own farm when converting to no-till, and that is why I address it so strenuously here. If your garden site is hard and compacted, simply building beds overtop it will not necessarily give you the results you are after—simply not tilling is not enough. Let me learn that lesson for you.

Remember that you have an opportunity here to prepare your garden for many years of success. In the beginning, that may require a single tillage or some other means of mechanical disturbance. Avoid getting bogged down by the terminology of "no-till." A well-executed mechanical tillage may ultimately allow you to better, and more rapidly, sequester carbon, build soil, and increase plant productivity.

Mechanical options for breaking up compaction abound, but I want to emphasize here that if you do not have experience working with plows, tillers, subsoilers, or whatever mechanical approach you choose, hire someone whom you trust to do this work, or stick with using a simple tool like a broadfork. A subsoiler is an extremely effective tool for larger plots, and for breaking up deep soil compaction, but it can also cause long-term compaction if used when the moisture level in the soil is too high.

Before performing any mechanical compaction mitigation, check to make sure the soil feels crumbly between your fingers and not at all slick or muddy at the lowest depth where you will be performing the tillage. Also have any cover crops or mulching materials on hand, because newly tilled soil is highly vulnerable to erosion. For larger plots, or when available, a subsoiler can be utilized if the compaction is deep. For clay-dominated soils, a few inches of compost applied to the surface then turned in with a tiller is the better approach to soil compaction mitigation. A half inch or so layer of sand from a creek (not a bag) can be added as well to help keep the soil from recompacting.

Any mechanical tillage should be complemented with biology by adding plants and an inoculating compost (one rich with microbial life), as described in chapter three. This is also a good time to add leaf mold for organic matter and to increase fungal activity. Breaking up compaction is an important starting point, but roots and biology must then fill the space created in order to provide air and space for new roots and microbes to move.

If you prefer not to use mechanical tillage in the process of decompacting soil, plants can serve the purpose. Note, however, that decompacting soil by planting a crop in it can require an entire season or longer. Or, if the

Clearing a Forest

When making a garden, clearing a forest is rarely an ideal situation but, increasingly, forested land can be the most affordable farmland available. If your plan is to start a market garden in primarily forested land, you will have to remove trees. Doing so will cause some ecological damage initially, but with proper soil management you can leave the area at least as diverse as you found it.

Keep these guidelines in mind:

- Use a topographical map to determine the portion of the site with the least slope.
- Clear a large enough area where trees to the south will not block too much sunlight.
- Hire a bulldozer to remove trees and stumps. It's safer and more efficient than clearing land by hand.
- If there are no large trees in the chosen area, look for a forestry mulching machine operator in your region. They use heavy-duty mowers that can grind small trees and shrubs into mulch.
- Leave some trees standing for shade and bird habitat.

compaction layer is deep or extensive, the decompacting process can take several years of repeated planting.

The most prominent example of breaking up soil compaction with plants is the use of long-rooted radishes called tillage radishes. This cover crop can be sown in the late summer or fall. The long roots of the radishes penetrate deeply into the soil then break down over the winter, opening up the soil to more air and better water movement. Forage radish (*Raphanus sativus* var. *niger*)—also known as the daikon or Japanese radish—is one type. Oilseed radish (*R. sativus* var. *oleiformis*) is another. Both types are considered tillage radishes, but there are some important differences between them.

Forage radishes generally winter-kill—that is, they die during winter after exposure to a few frosts. Oilseed radish also winter-kills, but it can be hardier than forage radish and thus is not ideal for growers in warmer climates. However, if either type of tillage radish does not winter-kill it can be killed with a low mowing in the late winter or spring. Tillage radish seed can be expensive and the roots are less effective at breaking up compaction in extremely wet and compacted clay soils. Tillage radishes can also bolt easily if sown too early in hotter climates. That early flowering will impede or eliminate the crop's ability to decompact. And if the plants are allowed to go to seed, that cover crop can morph into a weed problem. Even after beds are established, it is not a bad idea to mix these radishes in with a few other kinds of cover crops to break up any leftover or newly created compaction in the first few seasons. More about cover crop mixing and termination can be found in chapter seven.

The Never-Till Approach

If there is no need to till the soil—for example, if the organic matter content is high and compaction is a nonissue—you can consider options for preparing ground without much disturbance.

Where time allows, a long-season occultation can be utilized. This process, which is described in greater detail in chapter five, consists of securing a large sheet of opaque plastic (generally a silage tarp) overtop well-mown pasture to kill and break down the grass and to clear the surface. The surface must be well-mown or crimped (if it has a cover crop) before spreading the tarp, and the tarp must be well-secured to avoid being taken by the wind. Elimination of dense vegetation takes two months in the summer or three to five winter months, depending on your climate. Some warm, sunny weeks with temperatures above 70°F (21°C) will help stimulate germination of weed or grass seeds and will better break down root systems of aggressive perennials.

Figure 2.4. A few months before this photo was taken, this area was a pasture with perennial grasses, all of which died in the hot, suffocating conditions under the silage tarp.

An inversion of this idea is to mow the pasture low, form beds of compost, and then cover the beds with a tarp overwinter or for several months. Forming beds with compost is a common practice, but placing compost directly overtop a mown pasture may simply result in a newly invigorated pasture. The compost should be very deep in this situation to mitigate the risk of aggressive rhizomes establishing in the compost rather than dying. I prefer to rely on cardboard; using sheets of cardboard as a biodegradable weed mat stifles the pasture plants.

For this cardboard or paper mulch method, mow the vegetation as close to the soil surface as possible and then repeat the mowing a few times within a few weeks during the growing season. This intense mowing will weaken the root systems of most weeds, grasses, and perennials. Next, on a day with no wind, lay wet cardboard or mulch paper thickly overtop the mowed area, followed behind immediately by a layer of compost. Doing so will prevent the cardboard from blowing away in a breeze. This first application of compost to anchor the cardboard can be light, but ultimately the depth of the compost should be no less than eight inches. If you set up beds this way in the fall, plant a cereal rye cover crop directly into the compost to establish the beds and take advantage of the allelopathic (weed-inhibiting) properties of the rye plant. A tarp can also be thrown over the top of these beds and then removed in the spring before planting, though again, photosynthesis and living roots are always preferred.

Use tarps or a dense amount of carbonaceous material to kill the winter cover crops. Allow ample time for this vegetation to break down before planting cash crops.

Animal Tillage

Opening up soil doesn't always have to be a mechanical process. Some animals can provide a tillage-like effect while increasing soil fertility and microbial health. The two most popular animals for this are pigs or chickens; each has its strengths and weaknesses as a tilling "machine."

Pigs seem like the ideal candidate for soil prep, and they can effectively transform established pasture into bare ground in a short period of time. However, as pigs work through the soil with their strong snouts, they tend to mound the soil unevenly, which requires heavy work with hand tools to relevel afterward. Pigs also deposit all of their manure in one "bathroom" area and not evenly over the soil like ruminants or chickens. Being heavy animals, they may introduce some compaction, as well.

The less likely champion of animal tillage is the chicken. Long praised for their highly nitrogenous manure, chickens are excellent for building soil fertility, for consuming weed seed, and for slowly scratching soil open. A common approach is to house chickens in a mobile structure like the

Figure 2.5. Chickens working up a plot for future crop beds.

wheeled Chickshaw, as shown in figure 2.5, which can be moved every night to reduce manure buildup in any one spot. Portable electric fencing keeps the chickens contained and protects them from many kinds of predators.

Smaller ruminants such as sheep or goats can be used to mow areas very low, but leaving them on a plot long enough to mow grass down to the ground is nutritionally insufficient for the animals' good health and may result in parasite problems. Their weight and hoof action may also cause soil compaction.

In the past we have grazed cover crops with sheep, which can stimulate plant growth and add to the soil biology through their manure. Tierra Vida Farm in Colorado does something similar. In our conversation for *The No-Till Market Garden Podcast*, farmers Hana and Daniel Fulmer described how they sow new ground with cover crops, then graze it multiple times in a season with goats. This process helps Tierra Vida Farm bring up soil fertility and break up compaction in preparation for intensive vegetable production.

If you have livestock, consider their potential for preparing ground for crop production. That said, I do not recommend grazing ruminants as a technique for clearing the soil surface.

Transitioning to No-Till

Hannah and I farmed using a low-tillage approach for a few seasons before fully transitioning those gardens to no-till. In some ways, having established gardens made it easier for us to transition to no-till. We simply had to find a mulch we liked and eliminate our reliance on tillage for bed prep and crop turnover. However, in so doing we learned some big lessons.

The first lesson was that just because the soil is covered doesn't mean all the weed issues go away. Of course, most annual weeds are shallow germinators. If you bury them, and stop bringing them to the surface, they simply won't sprout. But perennial weeds store energy in their roots, so a simple mulch covering is not enough of a barrier to stop them from regrowing. We know now that if we want to transition a plot, or if we are setting up new beds in an area that has a lot of weed pressure, we should first use a tarp to occultate it for as long as possible. We then remove the tarp and pull out by the roots any perennial weeds we see. We place cardboard or paper mulch over the top of the weeded area and build the beds atop the cardboard.

One additional recommendation, though perhaps counterintuitive, is to sow or transplant your most intensive and valuable cash crops in the areas where there is perennial weed pressure. Here's why: If you sow or transplant a crop that needs very little attention (sweet potatoes or winter squash, for instance) into a plot that has a lot of perennial weed pressure,

Figure 2.6. After tarping this growing area, we spread out cardboard to block weeds and create a more fungal soil environment (which many perennial grasses do not like) as part of preparation for permanent beds. Notice the previously tarped soil (*left*) on which the cardboard is going.

you will likely not tend to it regularly enough, and the weeds will proliferate. Instead, in areas where noxious weeds have grown previously, plant an important cash crop—especially faster-growing crops like lettuce, radishes, or arugula. That will force you to stay on top of the weeds—pulling or cultivating them—until they eventually subside. We've almost entirely eliminated Johnson grass on our property using this approach in combination with the mulching practices described previously.

Compaction is very likely to exist in areas that have been farmed using a tillage system. As mentioned earlier in this chapter, we did not adequately address compaction issues when we first transitioned to no-till. We forged ahead and layered several inches of compost mulch onto our gardens. That mistake was a second lesson learned, and it destined us to a couple of additional years of dedicated broadforking. If I could do it over, I would have approached that transition by subsoiling, applying amendments, biochar, and biostimulants with an inoculating compost, and then tilling. Then I would have immediately covered the soil with a deep mulch and planted. Pursuing that process instead of simply covering our tilled beds with compost would have saved us from a lot of the broadforking we were stuck doing for two years.

You can expect a transition period of a year or so before you really start to see the benefits of no-till. You and your soil will need some time to adapt to the new system. Ceasing tillage does not mean the beneficial microbial communities and soil structure will instantly rebound. The soil organisms must first rebuild all the infrastructure in the soil. Good biological and structural recovery will require time, along with a lot of photosynthesis, compaction mitigation, and the application of some healthy microbial inoculants like composts and compost teas.

Designing Permanent Beds

The term *permanent beds* is often associated with raised garden beds that are surrounded by substantial borders made of wood, stone, or other materials. That's not necessarily what I'm referring to when I use the term *permanent bed*. There is nothing wrong with adding physical borders to garden beds, but for our purposes the term broadly refers to garden beds maintained in relatively the same place over a long span of time. This approach helps relegate compaction to the pathways and not the growing space. Permanent beds allow for long-term soil improvement. Because the soil structure never gets turned over using no-till practices, it never has to be reconstructed.

We created the permanent raised beds on our farm by using the rotary plow on our BCS walk-behind tractor, set to the depth of about four inches. Our beds are 100 feet (30 m) long and 30 inches (75 cm) wide and are separated by 14-inch-wide (35 cm) pathways. However, we are transitioning to beds that are 48 inches (120 cm) wide and 50 feet (15 m) long with 18-inch-wide (45 cm) pathways. The original 100-foot (30 m) beds with 30-inch (75 cm) width worked adequately on our farm though, in retrospect, we would have designed our gardens differently to begin with if we'd known then about all the options I describe in this chapter.

BED WIDTH

To my knowledge, the most common bed width in market gardens is 30 inches (75 cm). This is largely because most small-scale hand tools and walk-behind tractor implements are designed with this width in mind. However, simply because it is the most common width does not make it the best for everyone. The reason we originally set up our beds as 30 inches (75 cm) wide was to fit our cultivation equipment. I regret that decision now because bed width is one of the hardest things to change once a garden is established. And there are many reasons to dislike 30-inch-wide beds in a no-till system.

One reason is that we do not cultivate nearly as often as we used to, and so those precision cultivation tools (sized to 30 inches) are increasingly obsolete on our farm. The cultivation tools we still use are two sizes of stirrup hoes. Another reason is the considerable loss in growing space when you grow in

Focus

Terracing

Throughout history and throughout the world, people have tended crops on hillsides using terracing to keep soil in place. Spencer Rudolph of Sage Hill Ranch Gardens in Escondido, California, farms on a steep hillside that his father terraced with a bulldozer. Making terraces involves moving soil downslope and flattening it into "plateaus" that stair-step down the hillside. Spencer recommends hiring an experienced contractor who is comfortable working on slopes and can make the plateaus perfectly level. Spencer also suggests making wider plateaus rather than small terraces that hold only one or two beds, because the wider width makes setting up irrigation and building physical infrastructure, such as tunnels and trellising, much more straightforward.

Figure 2.7. The terraces at Sage Hill Ranch Gardens in Escondido, California.

30-inch-wide beds rather than wider beds, as shown in figure 2.8. A bed that is 30 inches (75 cm) by 100 feet (30 m) provides 250 square feet (23 sq m) of growing space. A bed that is 48 inches (120 cm) by 100 feet (30 m) provides 400 square feet (37 sq m). More growing space allows you to grow more crops, and with more crops comes more photosynthesis and more income potential. One extra row of lettuce on a 100-foot (30 m) bed can produce an extra 30 or 40 pounds (14 or 18 kg) of lettuce. And at seven dollars per pound—our current wholesale price—that is an extra $280 per planting. Thus, when you

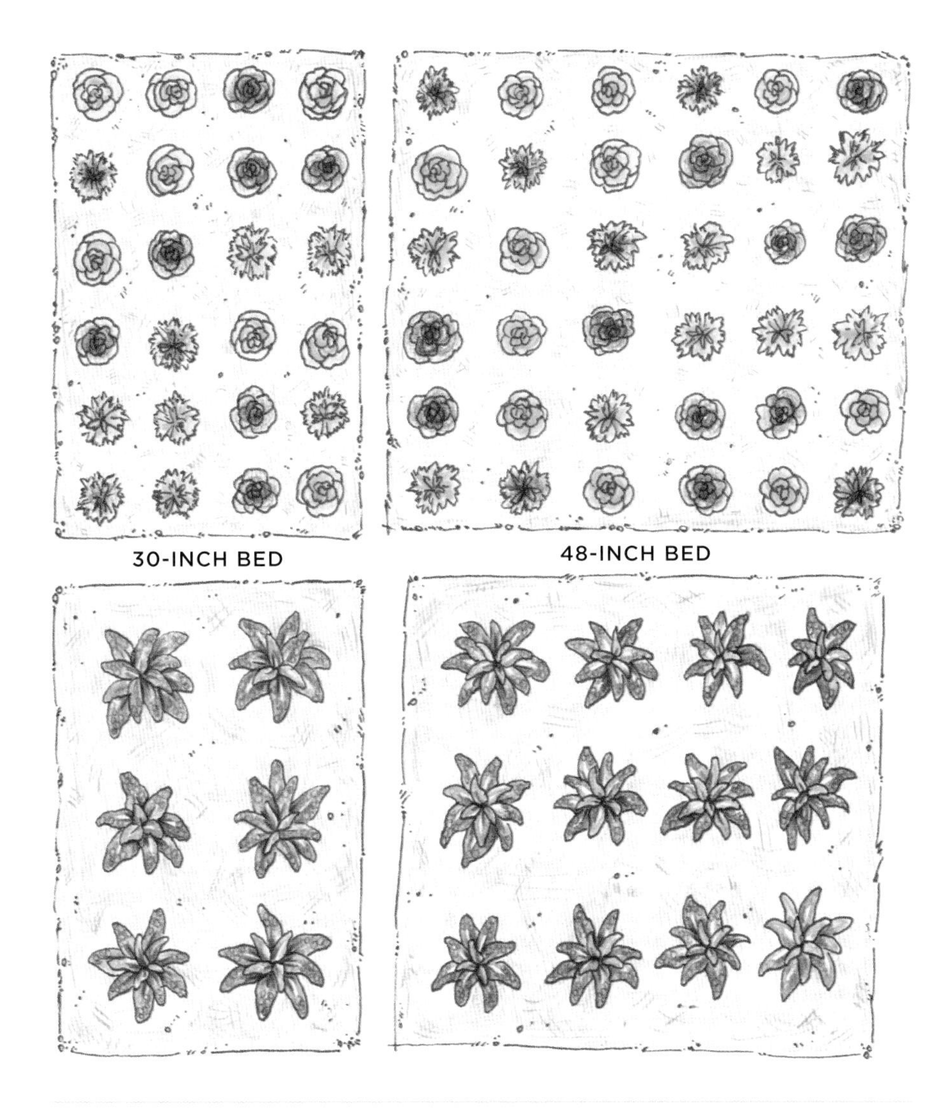

Figure 2.8. Wider beds can provide efficiency gains. On 30-inch beds, we usually plant two rows of kale. By adding just 18 inches, we can fit in four rows, doubling yield. With lettuce, we increase space between plants, and that boosts yield too.

don't have to rely on precision cultivation tools—most of which are sized to 30 inches (75 cm)—increasing bed width can make a lot of sense.

Furthering the case for a width other than 30 inches (75 cm) is that narrower beds mean more pathways to manage and keep weed-free. For example, consider a 50-foot-wide (15 m) garden plot. If you established 30-inch (75 cm) beds with 18-inch (45 cm) pathways, the result is 12 beds and 13 pathways to manage. Whereas, if you set up 48-inch (120 cm) beds and 18-inch (45 cm) pathways, you would have nine growing beds and 10 paths to manage. In that latter design you gain 20 percent more growing area—600 square feet (56 sq m) if the beds are 100 feet (30 m) long—with three fewer paths to manage in the same area.

Of course, 30-inch (75 cm) beds do have the advantage of being easier to harvest from or to walk over without stepping on and possibly compacting the garden bed itself. Compaction from occasional footsteps may not be an issue, however, depending on the no-till system you design. U.K. grower and author Charles Dowding, who popularized the deep compost mulch system (which he calls no-dig), regularly steps on his beds as he works in his gardens. We have begun stepping on our beds more frequently, too, because we find it packs the compost together. Even a shallow layer of compost mulch can sometimes create a growing area that is *too loose*, and so it does not hold the requisite moisture needed to germinate or grow crops effectively. In this case, occasionally stepping on beds can be beneficial.

Equipment can be a determining factor in bed width. If you intend to use a tractor, the spacing between the wheels (which can be altered on many models) will determine the bed width. To some extent the size of your tractors, and the implements used with them, will also determine how your beds are laid out because you will need to account for the amount of space needed to turn around at the ends of the beds. Chances are you will need an ample amount of space between the end of one section of beds and the start of another.

PATH WIDTH

Chapter six takes a dive deep into pathway management. But let's consider path width here, because it is an important consideration during the initial setup of garden beds. The more path space you have, the more nongrowing space you have to manage. Eighteen-inch pathways are more comfortable to work in than narrow pathways. But wider pathways require more labor to keep free of weeds. Pathways that are too narrow can cause issues when planting and harvesting, though. At Rough Draft Farmstead, our field pathways are roughly 14 inches (35 cm) wide, but those in the three high tunnels are closer to 12 inches (30 cm) wide. This is a little tight, but our harvest bins fit on these pathways and it creates less "blank space" to manage overall. If you plan to establish living pathways, by sowing a mixture of ryegrass and clover in the paths, for example, set your path width to match the width of your mower. (Living pathways are also discussed in detail in chapter six.)

BED LENGTH

The characteristics of your site may determine bed length—trees, buildings, or slopes will all have their say. No matter what length you decide on, standardize all your beds at that one length.

Figure 2.9. Standardizing bed length and width does not limit your growing possibilities—a wide range of crops can thrive in a garden with a standardized layout.

Hannah and I decided to set up 100-foot-long (30 m) beds in the field, and 50-foot-long (15 m) beds in the tunnels because that fits our landscape well. However, there is nothing wrong with beds that are 60 or 80 or 130 feet long (18 or 25 or 40 m). Base the decision on your context. I do highly recommend establishing a single length for all of your beds if possible so that all of your row cover, trellis material, or landscape fabric fits equally well over every bed (never too short or too long).

If every bed is standardized, it's also easier to make a garden crop plan for the season. This is a significant point because crop planning is already a complicated process, and having to adjust for the varying yields from a bed of one length versus another of a different length would create a massive logistical puzzle that would have to be reorganized every season. In standardized beds, the crop plan can largely stay static from year to year.

A shorter standard bed length has some advantages, including faster clean up and turnover to another crop. Shorter beds can also be prudent for succession planting and intercropping, or for running small crop trials. Managing crop trials, especially when they fail, can get messy and expensive in longer beds. And superficial though it may sound, don't underestimate the value of the morale boost that shorter beds can provide—on a per bed basis, they take less time to set up, clean up, and replant.

BED ORIENTATION

The orientation of a bed in relation to compass direction—east to west versus north to south—is less important than how its orientation affects water shedding or retention. For example, if you live in an arid climate where rainfall is low, it may be more important to orient your beds across the slope than up and down the slope. That will help them to capture what rainfall you do receive. If rainfall is often excessive in your climate—as it is here in Kentucky, with regular two- or three-inch rain events—make sure to orient any beds on slopes to encourage them to shed water, even if you garden on only a gentle slope. That may mean angling the beds slightly downhill or even aligning them directly up and down the slope to ensure that water can flow out of the pathways. The compaction inherent to walkways can hinder water penetration. If water cannot drain, it will fill the pathways and eventually rush over the top of the beds, thus eroding your mulches or soil.

One exception to considering the impact of cardinal directions is when you intend to use temporary tunnels such as caterpillar tunnels over the winter. Generally speaking, if the sidewall of the tunnel is facing the south, the tunnel will capture the most sunlight (this is especially true north of 40 degrees latitude). However, any structure will cast a shadow to the north that is double the height of the structure in the winter. This is something to keep in mind for planning winter production in the field or when building a high tunnel. Will you be able to make good use of the shaded area on the north side of the tunnel? Areas with high winds may generally require orienting beds into the prevailing winds, rather than broadside to the force of the wind, especially when temporary tunnels are involved.

RAISING BEDS

Raised beds are simply mounded garden beds that stand several inches above the soil surface; but raised beds are not the right approach for every grower and climate. To some degree all no-till methods cause the height of growing beds to increase over time. For the moment, though, I'm referring to the mounding of native soil to form growing beds, because if you raise the native soil, the addition of a mulch layer will make the beds even taller.

Generally speaking, the advantage of a raised bed is that it warms up and dries out faster. Consider whether that is something you may need for your garden in your climate and context for the full season. If you do need raised beds in the spring but not in the summer, consider raising your beds and then applying deep levels of wood chips to your pathways after the first spring plantings. Raising the pathways slightly closer to bed level will help retain moisture. Or maybe you could use landscape fabric or an occultation with tarps to help warm your beds in the spring and then remove them for the summer. Every gardening situation must be treated differently. Here are some other situations to consider:

Where to Use Raised Beds

- In areas where springs are cold and wet but summers are not too hot; the raised beds will warm up and dry out more rapidly in the spring
- In tropical and semitropical areas where torrential rainfalls are common and shedding excess water may be imperative to keep crops from drowning
- In dense soils that need improved drainage

Where Not to Use Raised Beds

- In warm climates with occasionally severe droughts; raising the beds in this climate may allow for earlier planting, but the raised beds can become too hot in the summer for many tender crops, and they may also shed moisture quickly in dry spells
- In dry climates with very low rainfall; raised beds lead to more rapid loss of moisture
- In large-scale cover crop systems managed with a roller crimper; in this method of no-till production (as described in chapter seven), it is important that the soil is level so as to ensure any roller crimping implement achieves good soil contact
- In windy areas where raising the beds may lead to more water evaporation, drying out of crops, and crop damage
- In sandy or volcanic soils that already drain quickly

Establishing No-Till Garden Beds

Once you have decided on the bed style, length, and width, it's time to establish the garden beds. There are many ways to accomplish this, and many nuances to consider in terms of sourcing compost (which I discuss more in chapter three) and other materials.

In this section I want to provide you with a few options for creating beds that are ready for long-term production and weed suppression. Bed establishment can be done at any point in the year. Ideally, it would take place in the fall so that a winter cover crop can be sown in the beds and allowed to grow and enrich the soil overwinter. That is not a requisite—growers do not always get to choose the time of year to start their gardens—but having some time for a crop to photosynthesize before putting the bed into market production is always beneficial.

Before filling in the bed areas with any mulches or crops, make sure to measure and mark the entire plot so that you know the beds are uniform in terms of width and length. There is no one right way do this task, just make sure that whatever you use to mark the beds is easily visible so that you can

Lowered Beds for Dry Climates

In arid and semidesert environments there are traditional gardening methods in which garden beds are lowered or protected with small borders to maximize moisture retention and reduce exposure. Latdekwi:we, or "waffle gardening," is a technique used by the indigenous A:shiwi (Zuni) people of New Mexico. Clay-based borders are formed around plots to direct any precipitation toward the plants. So named because of its wafflelike appearance, waffle gardening is generally a small-scale method, and intensive, but some potential exists for adopting this ancient tradition to meet the needs of market gardeners in regions where water is scarce.

In West Africa, a similar system, referred to as zai pits, has been widely employed for growing crops. In this case, large holes are dug and then plants are placed several inches below ground. Once established, the pits or "waffles" are replenished regularly with compost and organic matter. There is plenty of precedent from indigenous growers around the world for employing inverted beds in a desert market gardening context. Figure 2.10 is one conceptualization of such a garden.

Figure 2.10. Based on the zai pits system from West Africa, here's what a market garden in an arid region might look like.

clearly see the bed layout and not trip over the markers. I like to mark the edges of beds with string suspended about eight inches above ground level. I tie the strings onto short but visible stakes or fiberglass poles. Many growers pound wooden stakes at the ends of their beds to permanently mark the corners. A string is then tied between stakes from one end to the other, which precisely outlines the bed. This practice helps maintain the beds in the same place and at the same width season after season. When using mulch to form beds, that material can serve as the marker for where the beds are, but it is still a good idea to restring beds at least once a season to make sure you haven't inadvertently allowed them to shrink or become wider than the standard.

You can form raised beds by hand using a shovel, or you can borrow or buy a walk-behind tractor with a rotary plow. This latter implement tosses soil to the side mechanically, which establishes the path and bed at the same time (it is also an excellent implement for plowing up small fields). A smaller-scale version of the rotary plow is the Tillie, an electric tool that also tosses soil to the side. Heidi Choate and Evan Perkins of Small Axe Farm in Vermont use this tool to great effect. They farm on a relatively steep hillside with beds going across the slope. Gravity inevitably pulls loose soil downhill, and so they use the Tillie to toss the soil back uphill to replace the soil that comes off their garden beds every season. Ultimately, use whatever tool makes the most sense for your scale and budget.

After the beds are marked and the ground is prepared, lay out cardboard or paper mulch in the bed area. Then dump compost over the bed area and use a rake to spread it across the cardboard (there are also compost spreaders built for tractors that could be used for this step, just adjust the strings that mark the beds accordingly). The beds can be planted immediately with shallow-rooted crops or a cover crop where time allows. You can find more details on deep compost mulching in "The Deep Compost Mulch System" section on page 69.

Lasagna beds are formed using layers of carbonaceous materials and nitrogenous materials. An example of this would be to lay down a few inches of straw between the strings, cover that with compost until the straw is no longer visible, and then add another layer of straw and another layer of compost. A single layer of each is adequate, too. I describe the pros and cons of lasagna gardening in "The Lasagna Method of No-Till" on page 104.

Figure 2.11. At a 2020 No-Till Growers event at Jared's Real Food in San Diego, California, Jared Smith effortlessly pulls carrots from his lasagna beds.

The mulch created by mowing a cover crop is an excellent way to define and establish beds. Cover crops are selected based on season. An example is cereal rye planted in the fall to overwinter and then is terminated in time for late spring plantings. Another example is sorghum sudangrass sown in the summer to be terminated before planting fall crops or garlic. Regardless, the technique involves sowing cover crop seed thickly between the strings that mark the long sides of a bed, allowing it to grow for several weeks or months to the proper stage of maturity and then terminating it. Cover crops usage and termination techniques are covered in-depth in chapter seven.

It is possible to simply create your beds out of carbonaceous mulches, such as straw or hay. In this scenario, position the string higher than you would for other methods (or else the piled mulch will hide it). Shake a deep layer of the mulch on the surface to ensure no grasses come through. I recommend at least knee deep. Allow for several months of decomposition. This can be sped up by saturating the mulch with water, spraying with a compost tea, and then covering with a silage tarp. The material must remain well-saturated, however, to decompose adequately. Keeping the soil moist can be accomplished by adding drip tape beneath the tarp and running the water occasionally, or removing the tarp periodically to allow the material to become saturated through irrigation or rainfall. The pros and cons of all mulch materials are discussed more in the next chapter.

There are many no-till and low-tillage growers who do not use a mulch at all. In such systems, nothing is required to establish these beds beyond adding the amendments desired. Beds are amended, the top inch of the soil is raked or lightly harrowed to incorporate the amendments, and the bed is planted. In this system, each bed top should be delineated with strings.

Figure 2.12. This springtime view features beds sown with cover crops the preceding fall. The bed tops are still green and growing, and the pathways are shaded out, which helps prevent weed growth. When this rye crop is terminated, it will be easy to see the form of the beds.

PART TWO

Keep It Covered as Much as Possible

CHAPTER THREE

Compost in the No-Till Garden

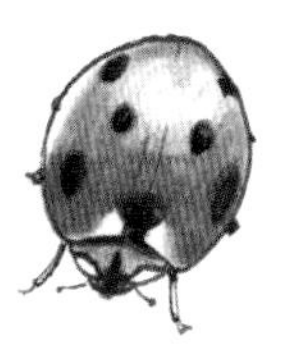

What is more emblematic of organic gardening than compost? Gardeners have added this decomposed organic material to their gardens since the first crops were cultivated to improve their soil and encourage crop success. Of course, much literature has been dedicated to the benefits of compost and the principles of making compost.

For many in the world of no-till market gardening, compost serves a very prominent role. For the purposes of stewarding living soil, compost is not just compost. Specially made composts inoculate the soil with diverse microbiology. Some composts serve solely as crop fertilizers while others are more suitable as mulches or as growing mediums. The use of compost is

Figure 3.1. We always have multiple piles of working compost on the farm to use as mulch, fertilizer, inoculant as needed.

an integral part of most of the techniques I cover in the rest of this book. So in this chapter I examine compost in-depth through the lens of the no-till grower, discern between four different types of compost, and describe how to work with each type.

To begin, let's define what compost is on a technical level. Agronomists tell us that any decomposed material is not technically a compost until its composition has reached a ratio of 10 parts carbon to 1 part nitrogen (the C:N ratio). If the ratio is any lower than 10:1, it's still manure or rotting plant matter. But all too often what is sold or used as compost does not meet that criteria. Compost is often composed of a wide variety of organic materials at varying levels of decomposition. For that reason, products distributed as compost can vary remarkably in C:N ratio, nutritional profile, texture, microbial makeup, and purity (that is, lack of contamination). Emphatically, not all composts are created equal.

Not all composts are intended for the same job, either. A composted chicken manure, for instance, should not be used as a surface mulch; whereas composted wood chips are not intended to address nitrogen deficiencies. All that to say, compost plays a very important role in your success as a steward of living soil, and so it's important that you know which type of compost you're dealing with, how to use it, and what to look for when shopping for compost, or making it yourself.

The Four Types of Compost

Because referring to all types of decomposed organic matter as "compost" doesn't give us the information we need, I distinguish four categories of compost: inoculating, fertilizing, nutritional, and mulching. This is not a common system of labeling composts, but it reflects the way I use compost and the qualities of compost that are most relevant to a market gardener. Each are used differently from the others, though there are situations in which you would use two or three different types together, as well. In this chapter I break out those details and work to offer you a more helpful way of looking at compost.

INOCULATING COMPOSTS

Inoculating composts are well-made, biologically active composts that are fine in texture. A compost like this can be used to inoculate an area that is deficient in beneficial microbial populations. The inoculation is done either through the use of compost teas or direct application of the compost to the soil. You may not have access to these types of composts commercially, but you can make them yourself in small batches as described in the "Simple Inoculating Compost Recipe" on page 61. Inoculating composts tend to be the most expensive type to buy. If shopping commercially, look for a professional compost producer—preferably an enthusiastic one. Ideally, you want someone who is looking at their compost through a microscope and discerning if it contains robust quantities of the right microbiology. In an ideal situation, the compost should contain everything from bacteria, archea, and fungi right through to amoeba and nematodes—biology the soil needs to convert organic materials into plant-available nutrients.

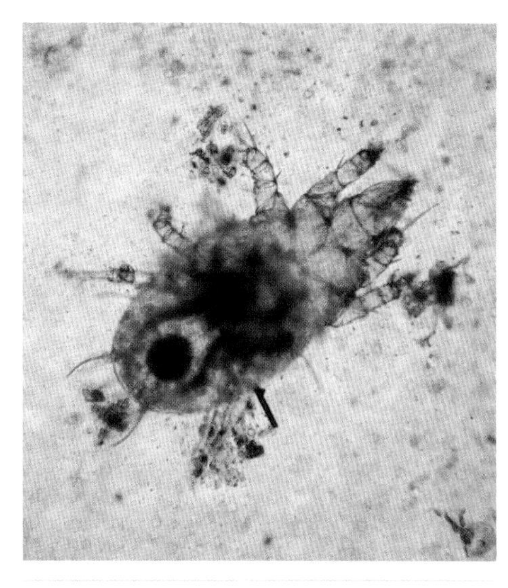

Figure 3.2. This "good guy" microarthropod was found in a homemade inoculating compost. Photo courtesy of Troy Hinke of Living Roots Compost Tea.

Vermicast—sometimes called vermicompost—is the "manure" of worms, and it is a prime example of an inoculating compost. Researchers have found that vermicast works well as a foliar application or when added directly to the soil. When vermicast is added to soils (or soil mixes) at rates between 5 and 40 percent, it promotes plant growth. However, researchers have also found that when more than 40 percent vermicast is added the results can be nonexistent or even negative. That's an important consideration. Tempting though the idea of a garden covered in vermicast may be, not only would that be wildly expensive, but it could even be harmful. Therefore, inoculating composts like vermicast are not recommended as a mulch. Instead, they should generally be applied before new crops go in the ground, or they should be mixed into nutritional composts as described next. On our farm, we almost always apply inoculating composts in extract or tea form rather than as a bulk material.

Simple Inoculating Compost Recipe

I worked with Troy Hinke of Living Roots Compost Tea to improve my inoculating compost recipe. Similarly, I highly recommend that you reach out to someone in your area who works with compost and compost tea for some guidance. Also note that the method I use was partially inspired by Master Cho and Korean Natural Farming.

You will need:

- A 5-gallon bucket filled with leaves, wood chips, and other forms of carbonaceous material such as straw, seed, nut hulls, or whatever is available locally

Figure 3.3. The initial carbon pile with a mixture of old grains, wood chips, and leaves.

- A couple of pinches of forest soil from below large, healthy deciduous trees
- Several gallons of water without chlorine or chloramine, preferably fresh rain water
- A quantity (3 or 4 gallons) of nitrogenous materials such as manure or green plant material (preferably a mix of different sources)
- A large supply (15 gallons) of wood chips

1. Combine the pinches of forest soil into two gallons of water and stir vigorously for several minutes. Pour the mixture over the leaves and other carbon materials in the bucket until covered and allow to soak for an hour or overnight. If you have to add or subtract a little extra water, that's okay.
2. Next, pour the carbonaceous materials onto a bare soil surface in a pile around 10 inches deep and cover with an opaque cloth sheet. I do this entire process under cover in a shed or other covered area to avoid rain.
3. Within a day, you should be able to feel heat from the pile. If the heat goes above 140°F (60°C), flatten the pile slightly or add a little water. Cover again with the sheet and leave until fungal hyphae—clusters of thin white hairs—are visible on the outside of the pile.
4. At that point, layer the wood chips, nitrogenous materials, and the fungi-rich material together in a lasagna-style pile. Start with some wood chips at the bottom, then add some nitrogenous materials, then add some of the fungal carbon. Repeat in layers. If the nitrogenous materials are heavily manure-based, the presence of wood chips will be highly important because they balance out the nitrogen and phosphorus. Wood chips or sticks also help add aeration to the pile and reduce the need for turning. The goal is roughly 30 parts carbon to 1 part nitrogen, but it does not have to be exact.
5. Add more water as you build the pile to moisten it. Your ingredients will determine the amount, but as a rule of thumb water should drip from

the material when you squeeze it but it should not feel like soft mud. To add the water, I find that the showering nozzle on a watering can or hose works well to saturate piles.

6. If the ingredients include any animal products, you are required to follow the rules of the Food Safety Modernization Act or the National Organic Program regarding compost materials, temperatures, and turning process. We prefer to use no animal products in our inoculating composts. That way, we do not have to turn the pile as much and can maintain slightly lower temperatures—between 110°F (43°C) and 140°F (60°C)—which are more favorable to fungi. These "vegan" inoculating composts have rendered excellent results as determined by microscopic examination.
7. If the pile smells unpleasant or simply pungent at all, cover the pile with more carbon.
8. Turn the pile fully one time after the first day. Keep it covered with the sheet.
9. If the temperature of the pile goes above 141°F (61°C), add a little extra water or flatten the pile out slightly. If the pile cools off, evenly flip the entire pile or add a bit more nitrogen. Think of this project like a soup you have to tend to and coax in the right direction.
10. Keep the pile moist and covered with a breathable opaque sheet throughout the process.

Figure 3.4. Even though this compost pile includes unshredded materials such as large green leaves, it will be fully transformed into nutrient-rich compost in just a few weeks.

11. The pile should heat up for several days. Check the temperature every day. Once it cools off, allow it to settle for a few weeks, maintaining moisture saturation so that water drips from the material is squeezed. After six weeks or so it is ready to be used in a compost extract or direct application. It can also be run through a vermicomposting worm bin to further enrich the compost. This is especially advisable if temperatures exceeded 140°F (60°C), a level at which many beneficial organisms can perish.

FERTILIZING COMPOSTS

Fertilizing composts are similar to inoculating composts in terms of their fine texture, but they are generally sought for their high inorganic nitrogen content—that is, nitrogen that can easily be taken up by plants. Composted poultry manures are the most common form of fertilizing compost available commercially; manure from rabbits, bats, and other small animals can be used for fertilizing composts, as well. Though the composition varies based on producer and other factors, poultry manures contain substantial amounts of ammonia, which rapidly gets converted by nitrifying bacteria into nitrates and ammonium—two forms plants readily use. Other material in a poultry manure-based compost is slower to release its nitrogen, meaning that it acts as a slow release "fertilizer" (more on nitrogen and fertilization in chapter seven).

Growers often add fertilizing composts lightly before planting to boost plant growth, but never as a mulch. The nutrient levels (nitrogen in particular) are too high to use as a mulch, and in some cases fertilizing composts can overload the soil or pollute groundwater when they are used beyond what is immediately required. Furthermore, its high ammonia content can become toxic to plant roots in excessive applications. Before applying a fertilizing compost, have your soil tested and consult an agronomist so you know what quantity to use. Never leave it directly on top of the soil, or much of the available nitrogen may be lost to evaporation. Lightly incorporate fertilizing composts into the garden beds using a rake or very shallow mechanical tillage. Or alternatively, cover them with a mulching compost.

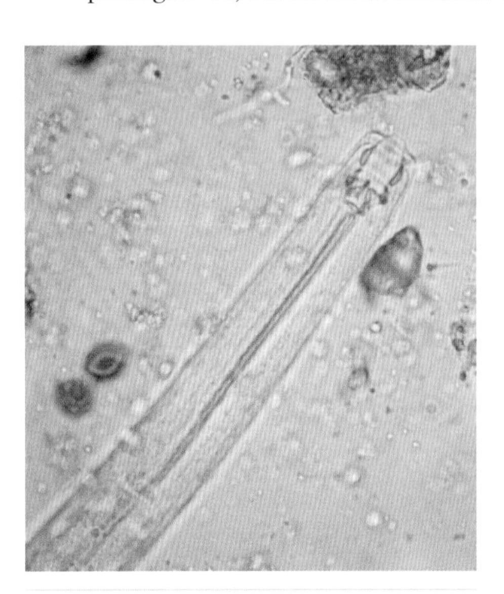

Figure 3.5. A predatory nematode, "good guys" found in this compost by Troy Hinke. Photo courtesy of Troy Hinke.

NUTRITIONAL COMPOSTS

Many forms of compost fall into the category of nutritional compost. Often made with some amount of animal manure—though not exclusively—nutritional composts are intended to add organic matter, microbiology, nutrients, and minerals all at once. Because nutrients are balanced out by the presence of ample amounts of carbonaceous materials—often wood chips or straw—nutritional composts can be used in greater bulk.

Some growers and compost producers pay close attention to process when making highly nutritious composts—fastidiously monitoring temperatures and microbial life, adding rock minerals, and so on. Other producers simply collect materials, follow any requisite composting guidelines, and sell the end product when it looks ready. Thus, the quality of a given nutritional compost is heavily dependent on a producer's choice of ingredients and their degree of enthusiasm for excellence. On our farm, we make our own nutritional composts using kitchen scraps and garden residues. We combine that with a little bit of an inoculating compost. The nutritional compost is added to the bed then covered with several inches of mulching compost as described below. A raised bed mix from a professional composter is a good example of a commercial nutritional compost.

MULCHING COMPOSTS

Compared to the other three types of compost, mulching composts are relatively high in carbon—20 parts carbon to nitrogen or more—but are not as high in nutrients. For this reason, it is safer to apply a mulching compost at

How to Find Good Compost

Finding good compost is a perennial job in market gardening. It can be a challenge for any grower but especially those in deeply rural areas or regions without an existent population of organic farms. Start by contacting other organic farmers in the region to get their input on compost suppliers. Landscape supply centers almost always sell mulching materials, and some will have decent compost for sale, too.

If you have a tractor or skidsteer, or don't need much compost, finding the raw materials to make your own composts can be easier than finding a good finished product. Expect that sourcing the best inputs to fit your system will take some time, and you may need to track down new sources—or double-check the quality of your existing sources—every year. The quality of your mulching materials will make up a large part of your garden success, so the effort is worth it.

high bulk because there is less risk of introducing excess nutrients into your garden environment.

Mulching composts are not finely textured like a vermicompost or a well-sifted nutritional compost. They tend to be less refined, often chunkier. This category of compost often includes mixtures of leaf mold, decomposed wood chips, composted hay or straw, vegetable scraps, mushroom substrate or mushroom compost, coffee grounds, and so on. Any carbonaceous organic waste product can in some way form a mulching compost.

These types of compost are used much like any mulch (as described in chapter four). Applications can be used around tomato plants or over the top of garlic. Or it can be applied to every bed in a deep compost mulch system, as long as you compensate for the compost's low nutrient load. If you intend to use mulching compost in a deep compost mulch system, you can use nutritional compost on the surface and then cover that with a few inches of mulching compost (I discuss this in more detail in "The Deep Compost Mulch System" section on page 69). Or you can directly amend a mulching compost with nutrients and minerals, as determined by a soil test. Testing and then amending enables you to better avoid nutrient excesses in the soil and around the plants.

Figure 3.6. A mulching compost often has a larger concentration of carbonaceous materials, like the wood chips and sticks visible here.

Risks with Compost

Please keep in mind when designing your no-till system that just because it's organic or free of chemicals doesn't mean that the system is incapable of polluting.

A primary consideration is the nutrient content of the compost, particularly the levels of phosphorus and nitrogen that some composts may contain. Excessive levels of these nutrients may not cause any immediate ill effects on crop growth. But when excess nitrogen and phosphorus escape the farm system (through runoff, for example), they can cause damage to aquatic life, contaminate drinking water, and create toxic algal blooms.

Composts made from manure are often heavy in phosphorus. Some growers contend that excess phosphorus in their soils is not a cause for concern. They've tested their water sources and have not found dangerous quantities of phosphorus in the water, and they have not seen ill effects in plant growth. However, the possibility still exists for excess phosphorus contained in compost to pollute, especially in heavy rainfall regions where soil particles containing phosphorus may be washed off the garden and into waterways. Excess phosphorus in soil has also been shown to inhibit mycorrhizal fungi.

Figure 3.7. Setting up new beds uses a large quantity of compost. To be safe in terms of nutrient overloading, we build the beds in a lasagna style with hay or straw on the bottom to help balance out excess nutrients from the several inches of compost in each bed.

Municipal Composts

Made from a wide variety of a community's organic waste materials, municipal composts are a common form of commercial compost. It's a challenge to categorize municipal compost as a nutritional, fertilizing, or mulching compost. Some municipalities produce a higher-quality product than others. Before purchasing compost from a municipality, inquire about their standards, how often they turn their piles, and if they allow sludge as an input (which is prohibited in organic production). In many states, compost producers are required to share their compost tests results. Ask the municipal composter to send you the analysis, and also ask what they do to mitigate contamination. Studying this information is the best way to know what you're working with nutritionally.

If there are no regulations for municipal compost in your state, proceed with caution. Ask other growers in your area about their experience with the local municipality's product. If no one else has used it, you probably don't want to be the first one to try it. Municipal composts may contain contaminants such as weed seeds, plastics, heavy metals, and persistent herbicides such as picloram and clopyralid (from treated hay or grass clippings). That said, some commercial compost may contain such contaminants, too.

The risk of phosphorus runoff needn't deter you from a system of deep compost mulch. I'm simply emphasizing the need to utilize compost ecologically and thoughtfully. Every compost will be different, depending on who made it and with what ingredients. So some composts do not have excessive nutrients, especially well-made nutritional composts. Other composts do, and they may contain other problematic contaminants, as well. I know several growers who recently suffered significant crop losses from composts that were contaminated with broadleaf herbicide. Some broadleaf herbicides, especially those containing picolinic acid, do not decompose readily in the composting process, therefore these persistent herbicides can easily damage vegetable, fruit, and flower crops or reduce yields. They enter the composting facility through manure, hay, straw, and lawn trimmings. To be safe, the US Composting Council recommends doing a growth test (called a bioassay). This organization provides details on their website for conducting a bioassay, but essentially you grow herbicide-sensitive crops such as peas or beans in the compost in cell trays and then look for the telltale signs of leaf curling and deformed leaf structure.[1] Ideally, that bioassay is done before any plants are put into the ground. I recommend buying compost that is made following the standards of the National Organic Program (NOP), and asking what the composter does for testing and if they do their own growth tests. Gather as much information about compost as possible before you buy or apply to ensure you do not run into contamination issues. You must do all you can to protect your garden and crops from bad compost.

The Deep Compost Mulch System

One of the most popular no-till management systems is known as deep compost mulching. Often referred to as no-dig in Europe, the deep compost mulch system consists of garden beds that are formed using four to eight inches of compost. That compost acts as both the primary growing medium and the mulch. Generally, the compost is applied once per year, though some practitioners prefer to apply a thin layer after every crop instead of all at once. In other words, the initial application is deep and forms the beds, while the subsequent applications simply maintain that depth.

To apply the compost, many growers simply fill wheelbarrows and dump out the contents on the beds. This is what we do at our farm, too. We then spread the compost out evenly with a rake. There are many types of

Figure 3.8. Crops grown on a deep compost mulch. These beds can be made and then planted within the same day.

compost spreaders for this job, as well, including attachments for walk-behind tractors. On a larger scale, a front-end loader can be filled with compost and slowly backed up over the plots while a couple of people shovel the compost onto the beds. This requires significant labor and a tractor. Growers Bryan and Shannon of Broadfork Farm in Nova Scotia use a clever self-loading compost spreader designed by Daniel Haartman of Community Machinery that both picks up the compost and deposits it. Essentially, this implement acts like a scoop that is mounted on the rear of the tractor. They can back into the compost pile, lift of a scoop of compost, and then drive it over the beds. A mechanism in the bottom of the scoop turns and allows the compost to fall out evenly onto the bed. It's a brilliant design that eliminates the need for multiple people to scoop or rake the compost.

In general, the deep compost mulch approach to gardening is exceedingly simple and often ideal for starting a new garden that is immediately ready for crop production. The compost makes an excellent weed block and its dark color helps warm the garden in the spring.

One negative aspect of a deep compost mulch system is that affordable compost is not always accessible, especially in rural areas. Compost quality will vary significantly based on region. The compost may not be fully broken down and may contain products not permitted in organic agriculture (like sludge). As I discussed previously, depending on who produces the compost and what inputs they use, commercial composts can contain persistent herbicides as well, which damage or kill plants. Some poorly made composts contain weed seeds, as well.

Another challenge is that compost is not always an ideal growing medium. For example, it can provide challenges for direct seeding if the compost is not well-decomposed or well-sifted, because it tends to dry out easily. In environments that receive regular heavy rainfalls, the compost may wash out. Moreover, the dark color can get exceedingly hot and cause plant roots to slow their growth, which is perhaps why you see this system less in warmer climates.

Overall, as long as these considerations are taken into account, the deep compost mulch system is one of the most straightforward and quickest-to-establish no-till systems. And, once it is set up you can modify it quite easily, adding cover crops and other mulches as needed. One farm that has done substantial work with deep compost mulches is Singing Frogs Farm in Sebastopol, California, which operates on about three acres. This is an excellent farm to follow. Farmers Paul and Elizabeth Kaiser have done numerous talks and seminars that can be found online. They also offer regular on-farm workshops where interested folks can spend the day on their farm working and learning.

CHAPTER FOUR

MULCH

If our goal as growers is to emulate Nature, then we cannot ignore the near ubiquity of mulch in natural ecosystems. It's the leaves on forest floor or the thatch in prairies. And these mulch materials are not there by accident. Nature has a great deal of incentive to keep its soils covered and, as you will see, so do growers.

In forests, pastures, and gardens alike, mulches:

- Help retain moisture by shielding the soil from wind and sunlight
- Distribute that moisture more evenly throughout the soil
- Provide habitat for beneficial organisms, both macro and micro—especially fungi
- Act as fodder for certain microbes called saprophytes that digest these carbonaceous materials and slowly release the mulch's nutrients in forms that plant roots can absorb
- Armor the soil against the impact of raindrops or footsteps, helping to prevent compaction
- Limit the germination of unwanted seeds

On our farm, covering the soil with mulch has resulted in more positive effects in our gardens than almost any other action we've taken. Fortunately or unfortunately, there is no lack of affordable discarded mulch material in our society. Leaves are raked from yards every fall. Food waste from cities is diverted from the landfill and turned into compost. In grain-growing areas, straw is cut and baled after the grain harvest. Tree trimmers shred branches pruned along power line right-of-ways and turn them into wood chips. In fact, almost every area generates bulk mulch materials that can be acquired affordably for application on garden beds.

In this chapter, I discuss the materials I've used for mulch in the past and those I work with today. I also describe systems some other excellent growers have utilized.

Your choices of mulching materials will largely depend on what is locally available. If deciduous trees are not common in your area, for instance, leaves won't likely be, either. But perhaps there are organic grain growers in your region, and if so then you will have access to affordable, herbicide-free straw bales—straw can be an exceptional mulch. Or perhaps it's hay. Or maybe it's cardboard. Or perhaps you won't rely on a mass supply of a particular material, but will work with a combination of mulches.

Compost has become a popular mulch and growing medium, but you won't find compost mentioned much in this chapter. That's because compost is covered in-depth elsewhere in this book (see chapter three), including the use of compost as mulch. Refer to chapter three if compost mulching is your desired approach, or if it's simply something you would like to occasionally incorporate.

No matter where you live, a little hunting should lead to at least one adequate form of organic cover material that your gardens will appreciate, though likely you will find several. In fact, I hope this chapter opens your eyes to all the potential mulches that exist. Let's review the best and most popular options, including the positive and negative attributes to each.

Straw

After a grain crop is harvested, the stalks of that grain can be cut, baled, and sold separately as straw. Almost any small grain crop will yield a straw byproduct. Common versions include wheat, oats, barley, and rice straw.

Often available in small square bales, straw is light and easy to spread. Because straw does not contain high concentrations of nutrients, overloading the soil is not a concern—in fact, being mostly carbonaceous, straw can help balance heavy nutrient loads from compost or manure when combined.

Straw can be used as a primary mulch applied to unplanted soil that is cleared of sod and weeds. It can also be spread as a mulch after planting. Straw mulches keep the soil cooler than compost mulch—that is one of straw's benefits in a warmer climate—and so the soil will be slower to reach adequate planting temperature if straw is the primary mulch. Couple it with an occultation or solarization treatment (I describe these methods in chapter five) to warm the soil slightly in the early months of the season.

In regions where grain is not grown, straw will be more expensive and less accessible. Here in Kentucky where few small grains are produced, a square bale of straw can cost between $7 and $12. Farther west in the United States where production of wheat, rice, and other small grains is common, straw may cost between $3 and $5 a bale. The difference isn't trivial to a market garden farming operation. If it takes five or six bales to

Slugs and Other Pests in Mulch

Mulches sometimes encourage plant pests such as slugs, voles, or inconvenient inhabitants like ground-dwelling insects, but it's a manageable situation. Though not all slugs damage crops, slugs are not something you want to pass along to customers in their lettuce mix. If slugs pop up in a mulched bed, there are several organically certified treatments one could use, though an effective alternative we've found is to slightly bury a plastic pint cup filled with beer near the problem area as a trap. The slugs crawl in and drown overnight—we've caught as many as 15 in one pint cup overnight. Space these cups every 5 to 10 feet (1.5 to 3 m), and make sure to check with your organic certifier about protocols to avoid breaking any rules.

Keeping ducks is a common suggestion for battling slugs, but I do not recommend them in a production context during the growing season. Ducks can do more damage to your crops than most slugs would, and their manure can greatly compromise food safety if they're in the garden during the growing season.

After harvest, if you still fear slugs are present on your lettuce or greens, soak the greens in water for 10 to 15 minutes to ensure the slugs release. Slugs do not like being submerged and will crawl to the top of the container, at which point you can simply remove them. Double check the produce before packaging, but the result should be slug-free produce.

Cardboard and other carbonaceous mulches can also attract ground hornets and various types of bees to nest. These are not bad critters to have around—many species are pollinators—though they can obviously be dangerous for people working near the nest. Remember, however, that these insects are often beneficial organisms, so do what you can to retain them and not harm them . . . or tick them off. If a nest becomes an issue and it's not possible to simply avoid the nest area, then destroy it safely (wear a bee suit).

Gophers and voles enjoy the presence of mulch, partially because the mulch can bring in food sources, such as worms. Increasing habitat for predators such as owls, hawks, snakes, and others can reduce the presence of ground-dwelling mammals. We keep farm cats to control rodents, though I should say that a deep mulch system just looks like one big litter box to a cat. Providing a dedicated and well-covered litter box seems to help.

mulch one bed, for instance, then a few dollars difference per bale adds up quickly over multiple beds.

Straw is not without drawbacks as mulch. The first is seed. If a baler that was used to bale seedy hay is not properly cleaned before baling a field of straw, that otherwise perfectly weed-free straw may end up contaminated. Or if a grain crop was harvested incorrectly, grew inconsistently, or was simply weedy, the straw from that crop may also carry unwanted seeds into your garden. One fall, we mulched a strawberry patch with hay that contained some wheat seed. Effectively we planted winter wheat among our strawberry planting, which made a mess of that patch in the spring. Before you buy, ask the supplier if they have any concerns about the presence of seeds. Also, devote some time to searching through the straw once it arrives—which we *did not* do that year before mulching our strawberries—because the presence of seeds will be evident.

Chemical residues on straw are a possible concern, as well. Some grain growers may apply postemergent herbicides for weed control in grain fields, and residues can remain on the straw. A grain grower may also spray a crop with herbicides in the final stages to expedite the drying down process; this treatment is called desiccation. Residues of persistent herbicides carried on straw mulch can damage vegetable crops. This practice may not be common in every region, but it is worth investigating before you purchase. Simply ask your supplier if the straw has been in contact with any herbicides recently. Explain that you intend to use it in a market garden. Most suppliers will give you an honest answer. However, it's also the case that many suppliers truly do not know the provenance of their straw supply. Buy certified organic straw where possible.

Hay

Hay is grass that is cut when it's green, allowed to dry, then baled. Nutritionally, hay is excellent for soil, because grass roots dig deep into healthy soil to mine minerals for growth. Hay mulch can provide a range of nutrients and possibly nitrogen, as well. If you can find a system for utilizing hay, at least from time to time, there can be major benefits for your soil and crops.

Hay is not without its complications as a garden mulch, however. As with straw, hay helps cool the soil. This can be an advantage in the summer, but it makes planting difficult and slow in the spring when the soil should be warm to germinate seeds and to avoid a shock to plant roots.

More so than with straw, grass seeds are an issue with hay mulch. It takes a very dedicated producer to cut the hay before seed heads have formed. It is possible to find a hay producer who will time cutting their hay specifically to avoid seed, but the bales will be more expensive than standard hay.

There is a persistent misconception that any grass seed contained in hay mulch gets buried and doesn't become a problem in gardens. Our experience demonstrates otherwise. In our humid environment with 50 or more inches (125 cm) of rainfall per year, grass seeds often germinate directly in the rotting hay layers, competing with our crops. Perhaps in drier regions, seeds would be less likely to germinate. Some growers wrap bales in plastic for several weeks or months to solarize them as a way to help reduce the

Figure 4.1. We often mulch our nightshades with hay to retain moisture through the year. Square hay bales differ in size depending on the source, but we account for at least six bales per 100-foot (30 m) bed for surface mulching.

weed populations. (Spoiled hay may have fewer weed seed problems than newly baled hay.)

Similar to straw, another concern is that some hay producers spray their fields with broadleaf herbicide to kill undesirable plants. These herbicides can be very persistent, and you would not want to plant any broadleaf crops grown in market gardens into soil or hay mulch contaminated with broadleaf herbicide.

So-called spoiled hay—hay that is partially rotten and no longer adequate as fodder—is often the most affordable hay, and we find it is some of the best to use as mulch because many of the weed seeds sprout or lose viability in the rotting process. Some

Figure 4.2. Hay is highly nutritious for soil but weed seeds hide in the small seed heads, visible here.

FOCUS
Straw or Hay Mulching System

In this method of no-till, a thick mulch of straw or hay is applied to soil that has already been cleared of sod. The mulch is then allowed to slowly break down. I generally recommend putting the hay down thick, just above knee-high. Even deeper can be better for straw, because it is faster to decompose. Do this application in the fall and allow the mulch to break down over the winter. This can also be accomplished over a few warm summer months.

This mulching system works best with transplants as opposed to directly seeded crops, but it can be adjusted to complement either. For transplanting, simply make small openings in the mulch and tuck the plants into the soil beneath. Then pull the mulch close to, or up against, the plants to retain moisture. For direct seeding, pull the mulch to the side of the bed to expose a planting row. I recommend applying at least a couple of inches of compost to the newly uncovered—but likely very healthy—soil as a surface mulch before sowing seeds. This will reduce any weed seed competition. That combination can be very effective.

NOTABLE VARIATIONS

Certain grasses that grow in salt marshes require salty conditions to germinate and thus their seeds are not as likely to sprout in garden soils. Yoko Takemura and Alex Carpenter of Assawaga Farm in Connecticut use this more reliably "weed-free" hay as a surface mulch to great success on small acreage. This could be a viable option for any grower located in a coastal area. Straw and hay mulches are also often used to retain moisture, but hay or straw can also be used to protect against torrential rainfalls. For instance, Kip Ritchey and Angelique Taylor of Smarter by Nature use hay as mulch in northern Florida where tropical storms and hurricanes are common. Also common in that area are extreme heat and occasional drought, so the hay mulch has many advantages in their context.

Figure 4.3. Organic straw mulch being applied at Urbavore Urban Farm in Kansas City, Missouri. The growers suggest using silage tarps over straw to help germinate then kill weed seeds. Photo courtesy of Brooke Salvaggio.

growers intentionally spoil hay for a year by allowing the bales to sit in the open field and rot. The trade-off with spoiled hay is that it's harder to spread by hand than pristine hay. Spoiled hay often sticks together in clumps that are covered in fungi. If the straw is really rotten, simply purchase a little more than you think you need and expect that it will to take a little longer to spread. The good news is that all that fungi is consuming the grass and releasing the nutrients in a plant-available form, so it will be worth the little extra work.

Fresh Hay, Haylage, and Grass Clippings

Green grasses and broadleaf plants tend to be very high in nutrients, which is plenty of incentive to find a way to use them as mulch. Plus, a green mulch should not carry the weed seed burden that so often accompanies hay.

Helen Atthowe and the late Carl Rosato of Woodleaf Farm, which is currently located in eastern Oregon, ran some mulching trials in a strip tillage system. (I describe strip tillage, and specifically Woodleaf Farm's use of it, in more detail in the "Strip Tillage" section on page 124.) One interesting discovery from the trials is significant here. Helen and Carl found that a four-inch layer of chopped clover and grass directly improved crop yields and production when it was freshly cut and applied in the fall. Paths were left in as living mulch.

In Germany, agricultural consultant Jan-Hendrik Cropp designed a similar system of transporting fresh chopped hay from ley fields—crop fields periodically taken out of production and planted into legumes and grasses—to production fields.

The process looks like this: Ley fields are cut high, just before the grasses and clover reach flowering stage, using a specific type of mower called a flail mower collector. This flail mower throws the hay directly into a manure spreader. The manure spreader is then used to spread the fresh hay two to four inches deep on cleared soil or soil covered by a crimped cover crop. Young transplants are planted immediately afterward. Cropp also uses haylage—hay that has been fermented in an aerobic environment (like sauerkraut)—as mulch. He recommends leaving haylage to sit on the soil surface for several days before planting the crop, because the acidity and off-gassing of very fresh haylage can harm plants. This method is mostly used by large-scale operations, but Cropp says there are growers who utilize these methods on areas less than an acre in size. Direct seeding of crops is not common with this method, but Jan-Hendrik says he and other growers are working on tools and techniques to accommodate direct seeding.

Grass clippings from lawns also can be highly nutritious for gardens. However, be sure you trust the source of the clippings and are confident they do not contain unwanted herbicides or pesticides.

Note that removing hay or clippings from one place and transporting them to another to use as mulch does remove nutrients from the source site. Therefore replenishing the nutrients of the hayfields or mulch-supplying plots is essential to maintaining that resource—livestock rotations and compost application are both effective approaches to maintaining the health of those areas. Remineralization—that is, adding specific minerals based on soil analyses—may also have some benefits.

Cardboard and Mulch Paper

An abundant waste resource, cardboard—or corrugated fiberboard, as it is technically called—is an exceptional mulching option. This is especially true when starting a no-till garden from scratch, or occasionally for flipping beds . . . or covering up mistakes.

In a study of chemicals in cardboard, the group Appropriate Technology Transfer for Rural Areas (ATTRA) found that cardboard is "relatively benign" so long as it is brown cardboard printed with black ink only. No glossy, waxed cardboard or boxes that have white coatings should be used. Also avoid produce boxes that could be pretreated with fungicides.

Figure 4.4. A fast way to create garden beds is to lay down a sheet of mulch paper such as this from WeedGuardPlus, then several inches of compost, followed by wood chips in the pathways. The paper mulch acts as a temporary weed-block below the compost.

Certified organic operations are required to heed these recommendations, but every grower should, too.

Also in organic certification, all tape must be removed from cardboard prior to application. To remove any tape, fully saturate the cardboard for a few hours. Most tape will come off easily. Several growers use cardboard on a small scale, but there are many who have utilized it on larger acreage. At one point in its long history, Seeds of Solidarity Farm in Massachusetts was using cardboard on two and a half acres. The farmers, Ricky Baruc and Deb Habib, say they generally do not remove tape before placing cardboard, preferring instead to simply rake off the residual tape the following year after the cardboard has decomposed.

To apply cardboard, lay the sheets thickly against the soil surface. We prefer to run overhead irrigation during the entire process (even if we get a little wet) to help saturate the sheets. Even small breezes can quickly blow cardboard out of place, and the water helps weigh it down. Wet cardboard also decomposes more quickly over time. If no sprinklers or hoses are available, presoak each sheet. While laying the cardboard, we add a light layer of compost to help hold it down. Once all the cardboard is in place over a bed, we then apply a thick layer of compost. Other nitrogenous materials such as fresh grass clippings or veggie scraps may also work as a mulch atop cardboard. If not applied thickly enough, however, wind will push these lighter materials off the cardboard, then work the cardboard itself off the soil. Securing the cardboard with sufficient weight will help prevent it from being blown all over the farm—a common complaint with cardboard. We have experimented with simply covering cardboard with a tarp—no compost or nitrogen-rich material—and leaving it to break down. But without consistent moisture being applied directly to the surface of the cardboard, the cardboard does not decompose under the tarp rapidly enough to be a viable approach. For that, I suggest laying drip tape irrigation overtop the cardboard but under the tarps and turning on the water for several hours periodically.

Paper mulches are another mulching option worth considering. Some products—especially those purchased as craft paper—may not be allowed in organic certification, so check with your certifier. One product that we've had good luck with is WeedGuardPlus, which is an approved paper mulch by the Organics Materials Review Institute (OMRI). This material is available in various widths and lengths, and it can be used as a replacement for plastic mulches on the surface of the soil or as a "weed mat" below deep mulches, especially composts. Mulching paper can be expensive: A 50-foot (15 m) roll can cost over $20, or over $100 for 500 feet (150 m). Use it strategically. The cost is reasonable when establishing a profitable or perennial crop, but it's a little too expensive to apply every year to annual garden beds.

Wood Chips, Sawdust, and Bark Mulch

Wood chips can be an abundant resource in areas rich with trees that need trimming. How much they can and should be used as a mulch, however, is a complicated question.

Layering wood chips in pathways is a common practice, as I describe in chapter six. But as a mulch atop garden beds, wood chips are not generally ideal, at least for intensive annual vegetable production. One argument against using wood chips as a bed mulch is that as they decompose the explosion of microbial populations temporarily ties up the plant-available nitrogen in the soil. This claim is overblown, but there is truth to it. Several research studies show that wood chip decomposition borrows nitrogen only from the soil area immediately in contact with the chips. However, many annual vegetables are shallow-rooted crops that would be adversely affected even by that degree of shallow nitrogen tie-up. (Deeper-rooted perennials are less negatively affected by wood chip mulches.) A disadvantage of wood chip mulch is that woody particles physically work their way into salad greens, which are not fun to eat. And the chips can promote slugs. Removing either the chips or the slugs from market crops can be a time-consuming challenge.

There are ways to successfully utilize wood chips on garden beds. Paul Gautschi, known for his Back to Eden Gardening movement, depicted in the film *Back to Eden*, uses wood chips as a garden mulch when paired with a layer of compost (below the chips) and a layer of manure (above). Those layers of nitrogenous material make the difference. Another option is to mix wood chips with raw manure or other high-nitrogen sources and allow it to partially compost before spreading it on bed tops. This mixture makes for a nice mulching compost. At the San Diego Seed Company, Brijette Peña uses an abundance of wood chips in a system similar to Gautschi's. When I visited Brijette with Josh Sattin in early 2020, I was surprised to learn that she had access to many wood chips in such an arid climate. Brijette said a local arborist dropped them off by the truckload. Brijette's story demonstrates that region doesn't necessarily limit your access to mulch materials. Sometimes, it's simply about who you know.

Notably, I haven't yet tried this method myself, though I do see the potential for adapting it to the market garden scale. What I like about wood chips, especially those from tree-trimming companies, is that they often contain a diversity of types of wood, including wood from small-diameter branches. Chips generated from this type of brushy growth are called ramial wood chips, and I discuss them in more detail in chapter six. If you have access to ramial wood chips, these small diameter, young growth chips can be excellent for increasing fungal activity.

Figure 4.5. A heap of wood chips delivered by a tree trimming company serves as good fungal habitat—and an awesome jungle gym.

Chips from tree trimmers also contain leaves and sometimes nuts—it's always good to have a diversity of ingredients in the garden. Where possible, avoid straight coniferous wood chips as mulch. Wood chips from tree species in the *Juglans* genus, particularly black walnut, contain small amounts of a compound called juglone that can inhibit the growth of some types of plants. If this is what you have access to, mix the walnut chips with other species to mitigate any risk of harming your intended crops.

Whereas wood chips are often available extremely cheaply or free, sawdust and bark mulch often come with a cost when they are sold either by landscape supply companies or lumber mills. Both sawdust and bark mulch compact more than wood chips, and they actually reflect water as opposed to absorb it (or allow it to infiltrate readily). As I noted previously, wood chips steal nitrogen only in the immediate zone where they touch the soil. Sawdust has greater surface area per volume—as finer materials always do—as compared to wood chips or bark mulch, and so it is going to have a greater extent of contact with the soil and therefore utilize more nitrogen in the decomposition process. If this is a resource you have access to, start with some small trials. Perhaps mix them into composts and let them age instead of using them directly as a mulch on their own to reduce risk of nitrogen tie-up in garden beds.

Leaves and Leaf Mold

The important thing to know about leaves and leaf mold is that if they're available to you, you should probably use them as mulch (or at least a compost ingredient). Leaves are loaded with nutrients. Once they make it to the soil, fungi—many of which live on and in the leaves—use an impressive array of enzymes to break down the woody materials that leaves are composed of, releasing the nutrients in a plant-available form. In that way, adding leaves is an excellent way to feed both the plants and soil life without the potential of adding weed seeds from sources such as hay.

If the leaf supply is fresh, it's generally best to shred them before adding them to the garden. When leaves are applied to a garden bed without first shredding them, they tend to "mat" against the soil, which can limit water penetration and air movement, as well as hinder the emergence of young plants. I witnessed this firsthand one year with garlic: A fairly thin layer of unshredded leaves restricted some of the garlic shoots from coming through. If a shredder is not available, a mower can do that work. Simply dress the beds with leaves. Allow them to dry slightly and mow overtop with a flail mower (also sometimes called a mulching mower) to shred them in place.

Leaf mold is decomposed leaves; it looks more or less like any other compost. In terms of water and air penetration, leaf mold is much easier on the soil than raw leaves and is rich with fungal life. Full of nutrients and minerals, but lower in phosphorus than composted manures, leaf mold can make an excellent surface mulch. It's also very useful as an organic matter builder in the early stages of starting a garden when it is turned into the soil with the compost and other amendments.

Peat Moss

Though not often applied as a mulch on its own in agriculture, peat moss is commonly used by landscapers to cover grass seed and as a component of mulching composts. Peat moss is also sometimes added to garden beds or other mulches to balance soil pH, change soil texture, or increase moisture retention. The moisture retention is a big part of peat's appeal as peat can hold 25 times its dry weight in water. For that reason, I think of peat as a

Figure 4.6. My son Further pushes a Landzie peat moss spreader to help me cover up cover crop seeds on a new plot in an effort to sow the seed without tillage. This method is a landscaper's trick and works best with grasses, clovers, and other small seeds.

sort of "mulch conditioner," because it adds moisture-holding capacity to drier mulches. I have had success adding a one-quarter-inch layer of peat as a moisture retainer just before seeding beds. It is especially good for slow-to-germinate crops like carrots that need consistent moisture for a week or more. Lovin' Mama Farm produces both vegetables and flowers in New York, and they use peat to great effect. Farmers Corinne Hansch and Matthew Leon mix peat moss from a nearby bog into deep compost mulched beds to improve their germination rates (which can be tricky in pure compost beds). To do this, they form the beds with compost, spread the peat overtop, and then use a power harrow to blend the peat with the compost.

Peat can be hard to spread by hand because it is so light and blows away easily. Some lawn care tools such as the Landzie Compost and Peat Moss Spreader—typically used by land care professionals to cover grass seed after sowing—help ease that loss. Such tools can be useful when spreading a light coating of peat moss as a mulch. Notably, the same procedure for covering lawn grass seed with peat works for sowing cover crops in a no-till system—landscaping and lawn care professionals can be excellent sources of no-till inspiration.

When people discuss the use of peat moss in agriculture, you'll generally hear two different opinions: one is that there is plenty of peat and we shouldn't worry about overusing it; the other is that peat bogs are an enormously valuable carbon sink and we should leave them alone. As with most controversies, there's some truth to both arguments: there are hundreds of millions of acres of peat moss on the planet, and peat bogs are also ecologically significant and thus need preservation.

Peat bogs are some of the most effective carbon sequestering habitats on our planet—storing upward of 500 billion metric tons of carbon, or double that of all the world's forests, according to United Nations Environment Programme (UNEP). For that reason, peat harvesting is highly regulated in many countries, and it is also paired with peat restoration in some countries such as Canada, which supplies around 25 percent of the world's commercial peat[1].

Coconut coir is often suggested as a more environmentally responsible alternative, but coir production also has environmental costs. Coconuts are often grown in poor, tropical soils that are low in nutrient content.[2] The byproducts of crops grown on poor soil should be returned to those soils as much as possible. In effect, peat and many peat alternatives are ecologically and environmentally expensive. There is no good answer here. An input into one natural system is almost always an output from another. I suggest researching what is locally available, and then supporting companies and distributors that care about and address the environmental costs of their products.

Synthetic Mulches

Plastic mulches and landscape fabrics are two synthetic options that are often used as mulch material because they have many of the same effects as the other mulches noted here. Though they obviously do not feed the soil as organic mulches do, synthetic mulches do retain moisture, warm the soil, and block weeds quite effectively.

I am not going to make any sort of endorsement of plastic mulches. They are typically single-use products that ultimately clog landfills. Landscape fabric is slightly more justifiable as it is reusable. Typically, once a bed is covered with landscape fabric and secured against the soil with yard staples (also referred to as lawn or garden staples), holes are punched through the plastic at the desired plant spacing (drip tape is usually established below the fabric for irrigation). Plants are then transplanted directly into the holes. The landscape fabric helps warm the soil, maintain the moisture, and reduce open surface area where weeds sprout.

I have long found landscape fabric most handy in our tunnels in the early spring—they create warmer soils, which are more biologically active and offer more nitrogen to plants. Landscape fabrics are also helpful in the first year on a new garden plot to knock back weed populations.

The negatives of these materials start with cost. Many plastic mulches cannot be reused and must be discarded at the end of the growing season.

Figure 4.7. Landscape fabrics come with a high price tag and a high environmental cost, but they are reusable and effective for warming the soil, blocking weeds, and helping retain moisture for hoophouse crops such as these early tomatoes.

That's a labor cost, a materials cost, and a waste-generated cost. Landscape fabrics are more expensive up front, but they can be used over and over again. Synthetic mulches flake off microplastic or even nanoplastic particles, and this is cause for concern. Recent studies show that these particles do have negative impacts on soil microbes.[3] Note that landscape fabrics are not typically made from polyethylene and thus do not contain phthalates—a known carcinogen and toxin to humans. This is described more in-depth in chapter five. Softer, more flexible plastic mulches do contain phthalates, and that is another issue that must be considered. A less urgent concern is that use of synthetic mulches may contribute to compaction in areas where heavy rains are common because the mulch can trap water at the surface, where it will then weigh down the soil.

After covering a bed with plastic mulch, it's crucial to bury the edges of the mulch to keep it in place. Removing and refolding the fabric is a messy, tedious job. Storage of these plastics can also provide a significant obstacle; even folded, they take up a substantial amount of room.

In the United States, organic regulations forbid leaving synthetic mulches on the soil surface past the end of the season in beds where crops are grown (at bed edges is generally okay). Some growers use biodegradable plastic mulches, but many organic certifiers do not allow their use.

Cover Crops

The classic definition of a cover crop is simply a crop that does not have a direct cash value but is grown to enhance soil and ecosystem health. The living roots help hold the soil in place, gather nutrients and encourage microbes, fight plant pathogens, add carbon, and more. But the aboveground biomass of certain cover crops, such as cereal rye in the spring or sorghum sudangrass in the fall also form substantial, weed-blocking mulches—often called "mulch in place."

One example of using cover crops as a mulch is when cover crop mixes—almost always *mixes* to maximize diversity—are grown to maturity and then crimped to the ground. The mechanism for crimping can be a roller with a chevron design like those developed by Jeff Moyer at the Rodale Institute, or something simpler. For the roller crimper, this implement is usually mounted in front of a tractor and filled with water to add weight. This design helps to pinch off the stems and the leaves, preventing the plants from accessing the water and nutrients in the soil while simultaneously laying the crop down as a mulch. On the smaller scale are growers such as Susana Lein of Salamander Springs Farm. Susana broadcasts dry bean seed into a standing crop of mature winter rye, then pulls a heavy

Figure 4.8. Where possible we like to flail mow our cover crop as a termination method but also to provide a nicely distributed mulch.

Figure 4.9. Roller crimpers are often mounted on the front of the tractor with a seed drill mounted in the back so as to crimp and plant at the same time.

wooden palette across the rye to crimp the stems over the seeds. I go into significantly more detail on cover crop management and termination in chapter seven.

Of course, it's possible to plan for purposeful termination of some crops through exposure to freezing temperatures. Such a practice provides a winter mulch. In my experience, cover crops that winter-kill do not provide as long-lasting a mulch as those that are terminated in the spring. However, there are other benefits from a winter-killed crop. The soil benefits from having that mulch as protection and microbial food all winter, and in the spring it's easy to simply rake aside the mulch and plant.

There may be types of mulches available to you that are not available to me where I farm in Kentucky. Seek out all the potential sources of mulch in your area. I believe that having a variety of mulches on hand makes everything about no-till easier, because it allows you to fit the right mulch to the right situation. A light-colored mulch could be used all summer before switching back to a darker compost mulch for the winter. Or when you transition from spring carrots to summer tomatoes, for instance, you may want to top that bed with a thick straw mulch that will suppress weeds through the dry months and into the fall. The topic of the next chapter is how to successfully make transitions from one crop to another, which is one of the most important elements to master in no-till gardening.

CHAPTER FIVE

Turning Over Beds

One of the more challenging questions in no-till market gardening is, "How do I plant one crop after another in the same spot without tillage?" At Rough Draft Farmstead, figuring out how to turn over beds without tillage was our biggest obstacle in transitioning to no-till. In fact, it has been the defining challenge in the movement as a whole. No-till market gardening couldn't truly take root until this step was figured out. And, largely, it has been.

When Hannah and I were tillage farmers, we relied on two methods to turn over a bed. One method was to mow and then till. We would mow a crop after it finished producing, then use a rototiller to blend the residue into the soil. We were especially likely to use this approach for weedy beds.

Because the soil needed time after tilling to digest the plant residues, which temporarily tied up the soil nitrogen, we would wait a couple of weeks before harrowing and replanting that area. This period incidentally served as a stale seedbedding—that is, allowing weeds to germinate over 10 or 14 days and then lightly cultivating them out before planting—that helped eliminate weeds, but having a bed out of production for two weeks meant no income was being generated. (I describe the strategic use of the stale seedbed technique for weed control in the "Carrots" section of chapter nine.) Another disadvantage of this fallow period is that the flow of root exudates into the soil stopped. The soil life was not being fed, and so it had to feed on itself.

The other approach we used was to yank out the plants, root and all—because we assumed that roots left in the soil would be hard to work around when replanting. Then we would amend, till, or harrow, and plant the bed with a new crop. This approach sped up the overall process, especially in beds not overwhelmed with weeds.

The reason we would till and then harrow is that bare soil compacts easily. So after removing a crop, we would often find that the beds were too

Figure 5.1. Fast bed turnover is a major benefit of eliminating tillage—the two middle beds here were full of tomato plants that were removed and replaced with lettuce within a few hours' time.

hard to plant into unless we did some mechanical tillage. There were instances when we could simply harrow the bed lightly to break up the compaction, but the beds were just generally not loose enough to immediately replant something. Because of that final tillage, in this method of bed turnover, the bed can be replanted in a day or two, or sometimes within a few hours. It's fast, but it's also pretty hard on the soil, because the act of ripping out crop roots extracts and exposes much of the carbon the previous crop sequestered. Then as the new crop grows, the bare soil easily recompacts, perpetuating the cycle. When it's time to change yet again to a new crop, a similar amend, till, and harrow procedure had to be done.

So what are the alternatives?

This is a critical question. Remembering the importance of organic matter, the fungal connections, the macro- and microbial habitat, and the general health of the soil—the bed flip is what it all comes down to with no-till gardening. Although growers have devised methods for flipping beds, there is always room for improvement in every system. Certain tools could be fashioned to speed up the many methods of turning over beds, and they are discussed in this chapter, and new techniques could be devised for using existing tools. I describe in detail each of the methods that I have tested at Rough Draft Farmstead, from the use of a mechanized flail mower to techniques that smother crops with large tarps to methods that require little more than a knife.

Having several different approaches to the no-till bed flip at your disposal makes your no-till system robust and adaptable. Terminating baby greens may require a different approach than, say, terminating eggplant or a cover crop. This is why it is important to develop a no-till tool kit, because relying on just one strategy to flip or prep a bed will likely not suffice. I can easily use a flail mower for lettuce and arugula, for instance, but I find that spinach lays too close to the ground to easily kill with a flail mower. I prefer to remove it by hand or hoe. So embrace the proverbial "tool kit." Having experience with a diversity of different no-till bed management techniques will present you with enough flexibility to succeed under practically any circumstance.

In this chapter, I describe all the methods I know to effectively transition from one crop to another without tillage. Then you can decide which best fit your circumstances. However, before we turn to the technical details, let's talk about how to protect the living soil when we flip a bed.

Maintaining Soil Health in a Bed Flip

Growers have widely adopted the term *bed flip* for the process of transitioning from one crop to another, but the term is a bit oxymoronic. For a no-till farmer who wants to build living soil, physically flipping *anything* upside down is not the goal. We're using the word *flip* in the less-common sense of the act of changing something from one state to another. During a bed flip, we change a bed from, for example, a crop of cabbage to a crop of carrots while disturbing the soil as little as possible, keeping it planted as much as possible, and keeping it covered as much as possible—our guiding no-till mantra.

There is a massive amount of room for interpretation and innovation within that framework. I discuss fertilization and bed prep in chapter seven, but here we'll focus on how those principles manifest themselves for no-till bed turnovers.

LET THE ROOTS LIE

Where possible, the roots should stay in the ground during a bed flip. Obviously, do not feel obliged to leave a crop's roots in the ground when they're the part you eat (carrots, sweet potatoes, and the like). But for crops from which the aboveground leaves or fruit are the cash harvest, do your best to leave the roots in the soil when the plants are done producing. That big mass of roots contains a substantial amount of carbon that will continue to offer some leftover root exudates to the soil microbial life for a time while you are establishing a new crop. Also, the root ball itself will slowly break down, in a way that does not tie up nitrogen to the extent it would if the roots were chopped and tilled in.

Oftentimes, depending on the plant and season, decomposition of a root ball can take several months or a year. That process aerates the soil, and the decomposing roots serve as fodder for some types of microbes and ultimately become stable organic matter in the soil. Again, in the carbon cycle, it's about having a balance of carbon going in and carbon going out for

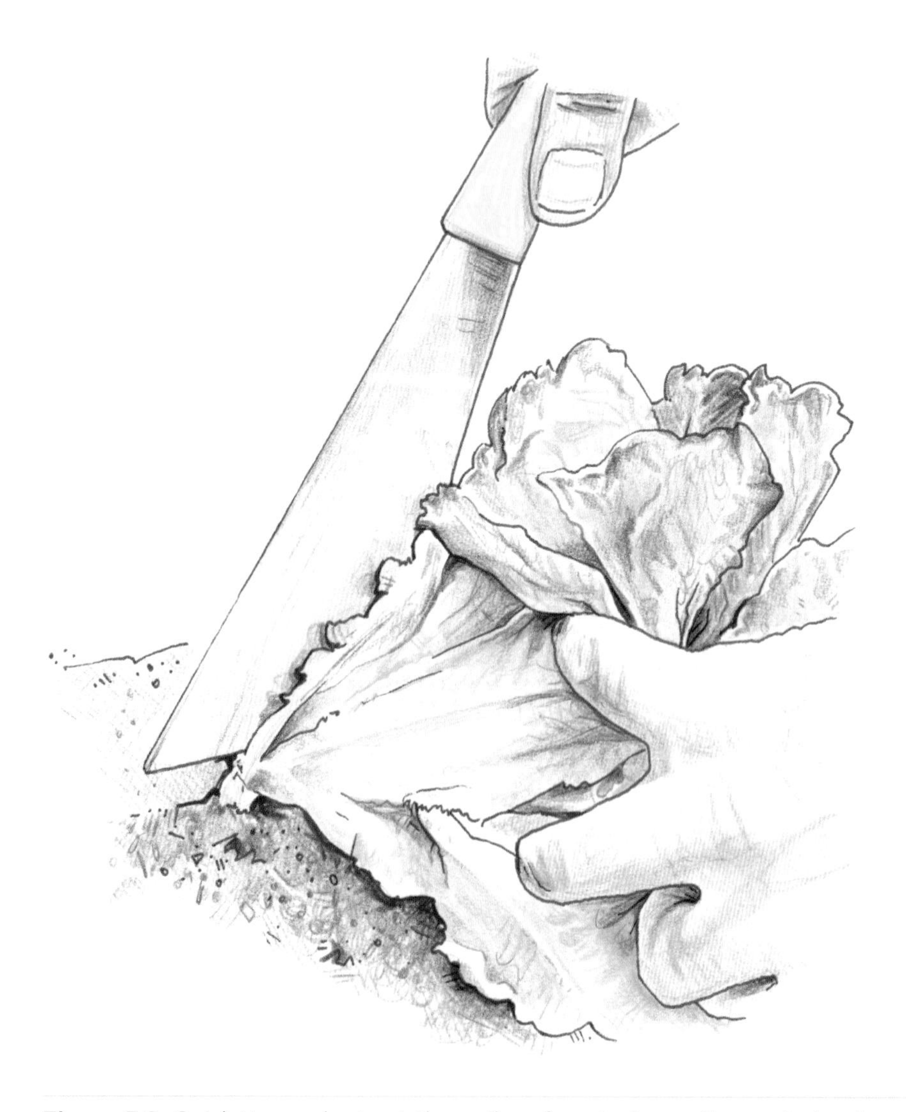

Figure 5.2. Cut lettuce plants at the soil surface to leave the roots in place. Some growers also twist out certain plants instead of cutting them to speed up the process.

proper nutrient cycling and microbial habitat. Leaving the roots in place lends itself well to satisfying this requisite of healthy soil.

THE CROP MUST DIE

The previous crop has to die before the new crop is planted, and some crops are easier to kill than others. Knowing which crops are difficult to kill (thick stemmed brassicas for example) versus those that die with relative ease (lettuce, spinach, arugula) is key. If the previous crop does not die, it will come back for a second life as a weed. Indeed, for a long time the biggest weed problem on our farm was whatever crop we had last grown.

Table 5.1 summarizes my experience of the best way to terminate specific crops. The results may vary depending on the crop variety and the time of year a crop is being cut—fall cabbages will generally winter-kill, for example, whereas a spring cabbage may readily regrow.

A handful of crops can be killed right at harvest. Others are terminated afterward. Regardless, it is good to know how to terminate crops with the least amount of disturbance possible. For some crops this involves cutting stems slightly below the soil surface; others can be cut just above.

When I say that a crop must be terminated below the soil surface, I mean that the stem of each plant should be cut systematically, one by one, at that location. I'm not talking about running a tool below the surface of an entire bed. Think of it more like surgery than bushwhacking.

Figure 5.3. When a plant such as this kale plant is removed from the soil roots and all, it takes with it a substantial mass of carbon and microbes that would be better left undisturbed.

Table 5.1. Crop Termination Methods

Crop Name	How to Terminate
Arugula	cut at surface
Baby Greens	cut at surface
Basil	cut slightly below surface
Beans	cut slightly below surface
Beets	remove root mass at harvest
Bok Choy	remove root mass at harvest; cut missed plants slightly below surface
Borage	cut at surface
Broccoli	cut slightly below surface
Carrots	remove root mass at harvest; cut missed plants slightly below surface
Cabbage (market/savoy)	cut slightly below surface
Celery	cut slightly below surface
Chicory/endive	remove root mass at harvest; cut missed plants slightly below surface
Cilantro	cut at surface
Collards	cut slightly below surface
Corn	cut slightly below surface
Cosmos	cut at surface
Cucumbers	cut at surface
Dill	cut at surface
Eggplant	cut at surface
Fennel	cut slightly below surface
Garlic	remove root mass at harvest
Kale	cut slightly below surface
Kohlrabi	remove root mass at harvest
Leeks	remove root mass at harvest
Lettuce	cut at surface
Mache	cut at surface
Marigolds	cut at surface
Melons	cut at surface
Mustard	cut slightly below surface
Nasturtiums	cut at surface
Okra	cut at surface
Onions (all types)	remove root mass at harvest
Parsley	cut slightly below surface
Peas (after pods)	cut at surface
Peppers	cut at surface
Potatoes	remove root mass at harvest
Pumpkins	cut at surface
Radishes (fresh and storage)	remove root mass at harvest; cut missed plants slightly below surface
Rutabaga	remove root mass at harvest
Spinach	cut slightly below surface
Squash	cut at surface
Sunflowers	cut at surface
Sweet Potatoes	remove root mass at harvest
Swiss Chard	cut slightly below surface
Tithonia	cut at surface
Tomatoes	cut at surface
Watermelons	cut at surface
Zinnia	cut at surface

AMEND AND COVER

Though optional in certain circumstances, refreshing the bed with a bit of fertility and mulch is a common part of the bed flip.

We do most of our mineral amending once per year before the season begins, but a small amount of amending accompanies most of our bed flips. We amend with organic nitrogen sources like feather meal, alfalfa meal, or fertilizing compost. We also add a small amount of humic acid, some kelp, an inoculating compost, and some biochar. Typically, the amendments are added to the surface and covered with a small amount of mulch, or they are lightly raked in if the existing mulch is sufficient. When we flail mow, we also add amendments beforehand. The mowing will then help cover the amendments with the plant matter. Incorporating amendments is discussed more in-depth in chapter seven.

Figure 5.4. Amending a bed as part of the bed flip process is not complicated. Using a five-gallon bucket full of well-mixed amendments, we simply disperse the amendments over the beds by hand.

REPLANT ASAP

Where possible, replant beds within a day of removing the previous crop. The goal of timely replanting is to allow the second crop to benefit from the previous crop's work of amassing diverse microbial populations and plant-available nutrients. In some cases we even plant the next crop into the existing crop before we harvest (see chapter eight for more on interplanting and relay cropping). It is always good to remember that photosynthesis feeds the soil and soil feeds the plants, so when there is no photosynthesis, the whole system is going hungry.

The Good and Bad of Occultation

Heretofore, this book has been dedicated to practices that maximize photosynthesis, but we're about to explore a technique that is nearly the opposite.

Popularized by Jean-Martin Fortier in his excellent book *The Market Gardener*, occultation is the use of an opaque sheet of some sort—a plastic silage tarp, generally—placed overtop a section of ground to kill crop residues after harvest is complete. The soil is then warmed by the cover and many of the weed seeds present are stimulated to germinate. These sprouts can't survive without access to air, water, and sunlight—can't survive, that is, without exactly the elements most plants need to grow—and thus they perish.

Figure 5.5. Large silage tarps can be used to clear plant matter, eliminate weed seeds, start new gardens, germinate crops, or terminate cover crops. Here, this tarp is in place to terminate a rye cover crop before tomatoes go in.

Restricting photosynthetic activity through occultation or "tarping" definitely has much potential for weed control and bed turnover. However, we can't neglect some of the negatives that accompany tarping, as well.

Occultation restricts moisture and thus allows growers to plant earlier in the spring or to prevent the soil from getting waterlogged during wet spells. Worms and other organisms may also "work" the soil below the tarps, balancing pH, eliminating pathogens, and enriching the soil with their dynamic digestive systems. Because of its passive nature, tarping is a very attractive option for no-till bed flips.

SILAGE TARPS

There are two main types of silage tarps—woven plastic and polyethylene black plastic sheets (usually 5 mil thick). They generally come in solid black or are reversible: black on one side, white on the other. Be sure that any tarp you select is UV-treated, which means the sun won't (immediately) degrade and weaken the plastic.

Most local ag-supply stores sell silage tarps. But shop around—prices vary. As for size, the largest that one person can, or at least should, reasonably move on their own is in the range of 2,600 square feet (242 sq m), such as a 25 × 100-foot (8 × 30 m) tarp (some suppliers provide 105-foot (32 m) lengths to account for complete coverage of the ends of a bed). Even moving a tarp of this size is a notably heavy job, especially if water has pooled on the surface. Larger tarps become completely unwieldy for even a couple of people to shift. We have found that cutting a 50 × 100-foot (15 × 30 m) tarp in half is fairly cost-effective, and it is easy to do on a relatively flat surface. Depending on whether a new silage tarp is rolled lengthwise or width-wise, you may be able to cut it with scissors while it's still folded. Otherwise, you will have to lay it out on a flat surface and cut it. This latter approach is generally the best way to go when cutting it to fit a certain width.

Tarps are spread over beds that have been mowed or are bare. If there is stubble left on the ground that you intend to cover, remove the stubble first with a low mowing or weed whacking. Any rigid plant stubble can poke through the plastic or lift the plastic, which may allow wind to more easily sneak beneath. Note that throwing tarps over standing grass will not achieve the desired effect, at least not very quickly. Silage tarps do rapidly break down plant matter that is not against the ground, but if a tarp is propped up at all on standing residue, grass, or weeds then it will not work. Such an area must be mowed or heavily crimped prior to adding the tarps.

Wind loves to play with tarps. It will happily take your silage tarp and toss it into your neighbor's field. For this reason, silage tarps must be well-secured. Sand bags are effective, as are cinder blocks. We use large

Used Billboards

Used billboards are extremely cheap, if not free, tarp alternatives for occultation. Be cautious, however. Although they will likely produce the same effect as silage tarps, billboards tend to be heavier. They also come in a variety of sizes. So piecing them together to fit a specific plot can be slow and tedious. Billboards may likewise bring with them the risks of compaction and also the risks from the leaching of PVC (polyvinyl chloride). PVC is an EPA-listed carcinogen with known toxins such as dioxin and hydrogen chlorides, as well as phthalates. Plastics containing PVCs are not allowed in organic production, thus ruling out the use of most billboards (since it would be difficult to prove that a billboard doesn't contain PVCs).

rocks because we have them in abundance and we can easily collect them in a wheelbarrow to redistribute as needed. We place them roughly 8 to 10 feet (2 to 3 m) apart around the perimeter, and we place a handful in the center of the tarp for added security.

In order to achieve effective weed control, leave tarps in place for a total of two weeks in peak summer, at least three to four weeks in the shoulder seasons, and a couple months over the winter.

Woven tarps are more expensive than sheet plastics, but because they are not made from the same type of plastic—that is, not polyethylene—they do not contain phthalates like sheet plastics often do. Phthalates are chemicals that are used in the production of many plastics to increase flexibility, but these chemicals can potentially leach into the soil. The issue here is that we do not currently have studies that tell us whether phthalates have negative effects on the soil or if they can be taken up by plants. We are starting to see some research on the negative effects phthalates have on human health, so that is where the concern is, because phthalates have been linked to neurodevelopment and behavioral issues, asthma, breast cancer, and more.[1]

So although no one is sure what the exposure levels of phthalates are after tarping, or what the level of plant uptake is, it's obvious that these are chemicals we would rather not be associated with as organic farmers or natural growers—folks who strive to produce the healthiest food possible.

Tarps also have the potential to shed microplastics—microscopic pieces of plastic that can potentially get lodged in soils or be consumed by soil organisms. These plastics can possibly affect soil health in the long term. A recent article published in the *Proceedings of the Royal Society B* showed dramatic impacts from microplastics on soil fauna, with effects that rippled all the way through the soil food web. Thus, if we do not consider the larger ecological impacts of our tools and methods, even when inconvenienced by them, we are doing no better than the types of agriculture we are trying to move away from.

Woven tarps allow for more oxygen and carbon dioxide exchange in the soil, creating a less anaerobic environment than is found under silage tarps made of slick plastics. Because moisture can pass through the fibers of a woven tarp, water typically will not pool as badly as on slick tarps. Though pooled water is helpful for holding tarps down, that pooling may cause some soil compaction over time on the polyethylene tarps. Compaction may then lead to negative anaerobic conditions as the tarps block oxygen and respiration. Thus, woven tarps offer the benefit of limiting that compaction and anaerobic activity, but they are both more expensive and heavier per square foot than slick plastic tarps.

LANDSCAPE FABRIC

Landscape fabric is a more compact version of a woven plastic silage tarp. The fabric is usually sold in long rolls that are about four or five feet (1.2 or 1.5 m) wide. The fabric can be used to occultate a single bed, as shown in figure 5.6. This can be particularly helpful if you want to occultate a bed in the middle of a plot that is too small to fit a silage tarp over. I was introduced to one effective bed flip using landscape fabrics or small tarps by Ray Tyler of Rose Creek Farms. Ray runs a very profitable and productive farm in southwestern Tennessee, and he has developed a method in which, for example, a bed of tender greens like arugula is covered with landscape fabric after harvest. He then saturates the bed with water, which will increase microbial activity and aid in decomposition. Then a sheet of landscape fabric is placed over the top. The microbes break down the previous crop residue, and within 5 days (longer during winter) the bed is ready to seed or plant without any other preparation.

It is also possible to plant directly through holes burned or cut into landscape fabric, and this tends to be an extremely effective option for weed reduction. It's essentially a way to occultate the soil while growing crops.

Whatever the kind of covering, this technique requires the use of giant sheets of nonbiodegradable plastic that someday will likely end up in a landfill or body of water. Environmentally and ecologically, these negatives can't be ignored, and that's why my goal is to avoid relying on smothering techniques over the long term. I am constantly working to reduce the amount of plastic I use by focusing on better crop planting. Keep in mind that as helpful as these tarps can be, certain weeds require sunlight to germinate. So warming the soil with an opaque covering may not eliminate all weed issues. Tarps can be highly effective but, as with any tool, occultation has its limitations. There are no perfect tools in farming.

Figure 5.6. At times on our farm, only a single bed in a large plot requires occultation. In this case we rolled out a 100-foot-long (30 m) strip of landscape fabric to occultate a single bed after a harvest of baby greens to prepare for carrots.

Though we use them less and less, we do still use silage tarps on our farm. They can fix a mistake, such as a poorly prepped carrot bed that gets too weedy and needs to be reseeded. Using silage tarps has enabled us to do trials we wouldn't have tried otherwise, such as perennial cover cropping. The tarps helped us to recover or control those plots after the trial without tillage. As you're developing your no-till system, tarps can be a handy tool. As your weed control and bed preparation practices improve, the tarp should slowly lose its place in your no-till system.

Silage tarps and landscape fabrics are permitted in organic certification. However, the National Organic Program standards state that plastic mulch must be removed at the end of the growing season. Some certifiers may be more particular than others about what constitutes the end of a growing season.

PLASTIC MULCHES

Another available "occultation while you grow" product is plastic mulch, which is available in both biodegradable and nonbiodegradable forms. Plastic mulches are often not reusable and are thus expensive—environmentally and monetarily—and I do not generally recommend them. That said, many growers have employed them strategically to great effect.

For example, grower Helen Atthowe made innovative use of black plastic mulch when she was farming in Montana prior to starting Woodleaf Farm with her late husband Carl Rosato. At the farm in Montana she used plastic mulch in beds but maintained living pathways. These pathways were kept in perpetual cover crop or "weed covers" (such as mallow, which is not a legume but can sequester 80 pounds (36 kg) of nitrogen per ton, according to Bob Parnes's book *Soil Fertility: A Guide to Organic and Inorganic Soil Amendments*). She would mow the pathways but plant into plastic mulches to grow beautiful field tomatoes and peppers in her northern climate. The plastic on the bed tops was effective at keeping the soil warm and reducing the "creep" of weeds from the pathways into the beds. There is more on living pathways in chapter six, but without question, Helen's methods could be adapted for a wide variety of climates.

As for biodegradable mulches, or "biofilms," these are generally made from at least a small percentage of material that cannot biodegrade in soil (though it may be as high as 90 percent non-biobased materials—that is, materials not derived from living or once living organisms). At the time of this writing, I am not aware of any biofilms that are free of inorganic binding agents and the like. Do your research before purchasing. A biofilm may eventually come on the market with a higher or complete percentage of biobased ingredients.

Plastic mulches are generally allowed in organic farming, though biofilms are generally not. Check with your certifier.

CARDBOARD AND PAPER ALTERNATIVES

I discussed cardboard and paper mulches in chapter four, but products such as WeedGuardPlus (or simply craft paper) have great potential in bed turnovers, as well. When a crop is finished producing, for instance, that bed can be mowed, covered with a light layer of wet cardboard or paper mulch, and then topped with a light or thick layer of compost. Then the bed can be replanted immediately or later. A drill with an auger bit is very effective for piercing holes through the weed-blocking layer, allowing you to insert plant roots through the barrier and into the soil beneath. Thus, you have the option of growing a crop while the weeds below are being blocked and killed by the cardboard. I recommend keeping the cardboard moist. That makes it easier to plant through if that is the immediate need, or it simply allows the cardboard to break down in time for when you are ready to plant.

An example of this bed flip method is the 2019 garlic crop at Rough Draft Farmstead. After the summer squash finished producing, we mowed the bed and layered cardboard over that area. We did not disturb the soil in any way. We then added a four-inch layer of compost over the cardboard and planted garlic. In part of the plot, we pierced the cardboard and planted the garlic in the native soil beneath. In the other part of the plot we did not pierce the cardboard and buried the garlic in the compost atop the cardboard. We did not see any noticeable difference in performance in this small trial. In both cases, the cardboard blocked the weeds enough to give us a perfectly clean bed of garlic in the spring. We have also used this technique with lettuce.

Weed Whacker, Knife, and Scything Bed Flips

I was first introduced to the weed whacker, or spin trimmer, bed flip by farmer Alex Ekins of Ace of Spades Farm in Spokane, Washington. Alex produces primarily baby greens, and he developed this system to make bed turnover fast and efficient. Using a bush blade on his weed whacker, Alex simply sweeps it back and forth across the bed, carefully trimming the plants at the soil surface. Those trimmings are then raked off or picked up, and the bed is replanted.

Like Alex, we found this technique to be fast and effective. The weed whacker acts almost like a scythe, cutting across a bed from right to left. This action sets the residue lightly and evenly on the bed, though it takes some practice to do it cleanly. It takes about 10 to 20 minutes to prep a 100-foot-long (30 m) bed for planting. When crops need to be clipped below the soil surface, we do sometimes use the weed whacker, but it can be messy work. If time allows, cutting plants individually with a knife or pair of loppers is preferred.

Figure 5.7. Garlic planted into cardboard and compost—the cardboard is still visible in the pathways.

Figure 5.8. A weed whacker is a versatile tool for bed flipping—in this case, clearing a bed of arugula.

FOCUS

The Lasagna Method of No-Till

In this style of bed formation, carbonaceous materials like straw or cardboard are layered together with a nitrogenous material like compost—each layer being well saturated as the beds are constructed. There is no set number of layers. These beds generally require some amount of time to decompose, but they eventually form mounded beds of excellent soil.

PROS AND CONS

The layers of carbon help to soften excessive nutrients in compost. The resulting beds are extremely loose and do not need much decompaction. In fact, deep lasagna beds are an effective way to circumvent tillage on lightly compacted soil. The weed suppression capabilities are very high.

Finding both quality compost and quality carbonaceous materials doubles the challenge of sourcing. There is a fair amount of labor involved in this process, and some amount of time must be given to allow the beds to "settle" so nutrients are not tied up while the microbes begin to consume the straw (letting the beds sit overwinter, planted to a cover crop, is ideal). As with any method that uses straw or compost, there is potential for contamination with weed seeds or persistent herbicides.

NOTABLE VARIATIONS

Lasagna beds are generally thought of as a small-scale permaculture method, but Jared Smith of Jared's Real Food maintains a very sophisticated system of lasagna bed production with straw and compost layered together on three acres in San Diego. Seeds of Solidarity Farm in Massachusetts builds lasagna beds of sorts using cardboard as their carbonaceous material. Briceland Forest Farm in Humboldt County, California, has had success using several *hugulkultur* beds—a type of lasagna bed that utilizes wood instead of, or alongside, light carbon materials such as straw—for cannabis production. At Rough Draft Farmstead, we build beds using hay and compost that is then tarped for several months to kill the weed seeds contained in the hay.

Figure 5.9. Logan Hailey and Cheezy Taylor of Ramblin' Farmers helping to build lasagna beds at a new plot where hay is buried beneath active compost. Paper mulch temporarily acts as a weed block in the pathways until the beds are tarped.

When you're starting out, a weed whacker with a bush blade is a more affordable option than a flail mower. It's also more efficient than a knife. I want to emphasize the use of brush or bush blade attachments, because using a plastic string cutting head that is standard on most weed whackers will result in shards of plastic being left behind in the soil. The bush blade is also very sturdy for woodier crops, which the plastic string will struggle with.

This weed whacker technique is very versatile in an intensive, small-scale no-till system. It's easy to carry a weed whacker into a tunnel, and it's easy to use to clear just a middle section of a bed or to terminate an interplanted crop. All of these tasks can be difficult to do with a flail mower. A weed whacker is, of course, a noisy tool and working with it can be physically exhausting. There are also electric weed whackers available if you would like to avoid fossil fuel–dependent machinery—Deb and Ricky of Seeds of Solidarity Farm use solar power to charge their house, and so they effectively charge their electric weed whacker with solar power.

SCYTHE TECHNIQUES

As improved agricultural practices arise, they present an opportunity to weigh the new utility of old tools. The scythe is definitely one tool that deserves a second look.

Scything is quiet. You do not need to put in earplugs to use a scythe. You don't need fossil fuels. You don't suffer mechanical gyration of your bones and muscles. There is something unquestionably soothing about scything.

Because crop termination requires cutting at least right at the soil surface, you will want a bush blade for flipping beds. A bush blade is slightly heavier, shorter, and more rugged than a grass blade. So if the bush blade hits the soil or a woody plant, it is not as likely to dent. A new bush scythe with the snath (handle) will run you about $200 to $300. Cruise the antique malls first, however, as an expensive precision scythe is not required for flipping beds. You want something you can ding up a bit, at least while you're learning.

The scythe works best for crops that are cut above the surface, as opposed to those that need to be terminated below surface (see table 5.1, "Crop Termination Methods").

As for technique, watch a few videos on scything. Scything is a task that can be very relaxing or it can wear you out rapidly, so developing good form is important. Figuring out a scything method depends on whether you have raised beds, whether they are sufficiently mulched (so that you can walk on the bed tops), and how wide the beds are. Once the plant material is cut, rake debris, pick any weeds by hand, amend, and replant. Jared Smith of

Figure 5.10. To cut top growth with a scythe, try to stand in the paths and cut one side of the bed at a time. It takes practice to find a comfortable rhythm.

Jared's Real Food scythes a bed of greens to remove the dense top growth. He then rakes off the cut plant matter and uses a wheel hoe to remove the remaining stubble. That's an effective technique to have in your toolbox.

KNIVES AND OTHER HAND TOOLS

Whether the outgoing crop requires being cut below surface or being cut at the surface level, a team of people can quickly flip a bed using little more than a long-bladed knife. Even if this is not your primary approach for flipping a bed, it is a good technique to know and have at your disposal. Garden shears and long-handled loppers both work well for removing thick-stemmed crops both above and below ground.

The machete is a cheap, efficacious, and enjoyable tool to use for turning over beds. It is extremely handy for cutting out woody crops, in particular thick brassica plants or field tomatoes and peppers that are supported using the Florida weave (this is shown in figure 9.23 on page 235). Among my favorite tasks every year is using the machete to cut the string from T-posts and then the base of every tomato plant at the end of the growing season. I've also found it effective (though a bit messy) for terminating crops such as chard or kale that need belowground cuts.

Figure 5.11. This lettuce field knife available from Johnny's Selected Seeds can be easily plunged into the soil to terminate difficult crops.

Mowing Methods

Whatever mower you use, always keep in mind that the bed must be cleared of weeds before mowing, as the vast majority of garden weeds are not going to die unless they are clipped below the surface or removed entirely.

FLAIL MOWERS

Unlike other mowers, flail mowers spin on a vertical plane as they cut, dropping the residue directly down as opposed to throwing clippings off to the side. We use the Berta 30-inch flail mower attached to a BCS 749, 13 hp. The Berta is a robust implement capable of chopping through very dense vegetation. There are other brands of push flail mowers, but I do not have experience with them. The Berta-BCS combination is not cheap—the pair may cost as much as $6,000 brand new. Walk-behind tractors like the BCS can carry out many other tasks, too, when paired with other implements: chipper-shredders, harrows, mowers, balers, and so on. If using the flail mower for bed flips is the only implement you want for your walk-behind, then I suggest you crunch some numbers to figure out how many hours of labor (thus money) this would have to save you.

How *many* hours can the flail mower and walk-behind save you? Using our farm as an example, it takes me 1 to 2 minutes to clear a 100 × 30-foot (30 × 9 m) bed of lettuce (about six hundred plants) with the slightly lowered flail mower. Even if the bed needs raking afterward, that bed can be ready to replant within 5 minutes. Removing the crop residue from that same bed by hand may take one person 20 minutes or more, even working at a good clip. Using a weed whacker, the task would likely take 8 to 10 minutes. We flip 30 or more beds of greens alone during a season, so that's

Figure 5.12. Shown here on a BCS 749, the flail mower is a simple implement that can speed up bed flips tremendously.

a significant time savings: 30 to 90 minutes of labor using the flail mower as compared to 10 to 15 *hours* of hand labor per season. And I'm only talking lettuce and arugula, not the squash, cover crops, or other plants that the flail mower method can be used for. Do your own math for your context, and you may discover a highly significant amount of savings is possible by switching to this technique.

In practice, we remove any weeds that might survive the mowing. Next we mow, then amend, then rake if needed, preferably with a tine rake weeder. (A bed rake with stiff tines may also pull out roots.) Then we replant. We have been experimenting with adding amendments before we mow so as to bury them beneath the residue, but you should do this only if you do not intend to rake (you don't want to rake the amendments off the bed). Also, if you do not have weeds under control, don't rely on the mowing method to terminate weeds for you—especially grasses that regenerate easily. See "Starting from Scratch" on page 35 for more details on getting your weeds in check from the beginning.

PUSH MOWERS WITH BAGS

Using an electric or gas-powered push mower with a bag allows you to pick up the spent plant matter from the bed top as you go. That plant matter can then be taken straight to the compost pile or spread on another bed as mulch. You may run into problems with the dense plant residue clogging

Modifying the Flail

Essential to the flail mower bed flip technique is lowering the blade. To do this, you should adjust the guides that determine cutting height—generally skis or a rolling bar that can moved up or down. Depending on how your flail mower is set up, level the rolling bar to be flush with the blades, like you would for a tight mowing (think putting green on a golf course tight).

This may take some dialing in. Simply lower the blade, test it, and keep adjusting it until you're happy with the result. The goal is to cut as close to the surface as possible, but without cutting into the soil. Some mulch will inevitably get caught up in the blades, and the mower will occasionally clip uneven soil surfaces. The majority of the soil, however, should be left undisturbed.

Figure 5.13. Lowering blades nearly flush with the roller results in the lowest cut possible—a quarter to a half inch (6 to 13 mm) above the soil.

the mower, as these mowers are traditionally intended to cut already shortened grass. A more powerful push-mower or mulching mower should be utilized for this—something built for more than simple lawn care.

There are types of flail mowers—generally called flail mower collectors—that both mow and remove the plant residue. I am not aware of such an implement designed for use with a walk-behind tractor, although there is a small baler for the walk-behind that could serve a similar purpose. Using a collector on a flail mower would replace the task of raking the bed after mowing. You could then use the collected plant matter elsewhere—for composting, mulching another bed, creating haylage, and so on.

BUSH HOGS AND OTHER MOWERS

Any mower that pushes debris off to the side could be fashioned for use in flipping beds, though you would likely not want a mower that pitches debris onto the next bed. That could have devastating effects on sensitive crops such as lettuce or arugula, as well as on any crops with tender greens such as turnips and radishes. A modified mower that could lay the residue from a bed into a pathway, however, could be extremely helpful for bed flipping and path mulching.

Stirrup and Wheel Hoes

Sometimes referred to as a hula hoe or scuffle hoe, a stirrup hoe is a standard garden tool that I highly recommend having on hand. With its swiveling stirrup-shaped, double-sided blade, this tool comes in handy for pathway management and also dealing with any weed outbreaks. And, of course, for terminating crops.

We have used stirrup hoes to clear whole beds, carefully slicing through the roots of the crop before raking the debris away. This work, however, requires no small amount of labor and it does disturb the soil significantly. We find that the stirrup hoe is a handy tool for clearing around interplanted crops. Let's say that the remains of a lettuce crop needs to be removed from among sweet pepper plants. The stirrup hoe works well to surgically remove each lettuce plant, causing very little injury to the soil or the pepper plants in the process. And you can use this tool standing up—unlike using a knife, there's no need to bend over to clear away the plants.

Wheel hoes are also excellent tools, designed to reduce body stress and speed up the cultivation process. Often they come with interchangeable attachments such as cultivating shoes, hillers, furrowers, and so on. Because a wheel hoe usually has a sizable price tag attached to it, if you purchase a wheel hoe new for bed flips and general cultivation, then consider how else you can put it to use, such as for path management or hilling potatoes.

Many low- and no-till growers use the wheel hoe to remove crops in densely planted situations. Baby greens are the prime example. Pushing the sharp stirrup beneath the plants right at the root can effectively terminate them. Then comes a raking, and the crop is removed. The depth at which it cuts can be adjusted at the handles, on the implement attached, or simply by the user.

Note that a wheel hoe may become clogged when cutting woody material such as cauliflower stems, and it is perhaps best used for tenderer crops. Also allowing the wheel hoe to cut too deeply into the soil may create compaction layers. This is especially true if the soil is dense and worked too wet (denser soils compact more readily than sandier soils—picture smudging a block of clay between your fingers). As discussed in chapter one, these compaction layers will limit water penetration, air movement, and nutrient access for roots and microbes. Always keep the blade sharp to reduce compaction potential.

Solarization

Solarization is a method of using a clear covering to amplify the sun's intense UV radiation to heat the soil. This technique has been popular for decades as a way to kill weed seed and disease organisms in the top several inches of a bed or field, but recently has become popular among growers as a great way to kill crop residue without any physical soil disturbance. It is also popular because it provides an excellent "second life" for used plastic film from high tunnels.

Popularized by farmer Bryan O'Hara, solarization is effective for flipping beds from one crop to the next. In his excellent book *No-Till Intensive Vegetable Culture*, O'Hara describes his experiences using solarization with great details on duration and proper usage. Following are the basics of solarization, combining some of Bryan's experience with our own trials at Rough Draft Farmstead.

Solarization and occultation are not the same technique, although there are some similarities. Occultation with opaque tarps can warm the air and soil underneath, but solarization with clear plastic sheets results in more extreme temperatures. On a hot sunny day, the temperature just underneath a sheet of clear plastic could be upward of 125°F (50°C), which is well above a reasonable temperature for photosynthesis in most plants. In the same conditions, the temperature under a silage tarp registers about 110°F (43°C).

Unlike an occultation using a black tarp to clear plant residue—which could take several weeks depending on the time of year—plants may be killed in just a few days or even in a single day when covered by a clear plastic sheet, depending on the crop type, sunlight intensity, and day length. On a sunny day, solarization is definitely a faster termination option than occultation.

That faster turn-around time with solarization means that it is less damaging to microbial populations, as it allows the potential for crop roots to hold on to more of the microbial populations in the root zone. When a bed has spent two weeks under black plastic, by contrast, the residual exudates would be limited, and thus microbial populations are more likely to suffer stress and decline. Also, as long as the duration of solarization is no more than a few days, the heat trapped by the clear plastic is not likely to penetrate too deeply into the soil, and microbes below the top few inches are likely to survive. Soil in itself is a great insulator, especially if also covered in a mulch. Thus, a bed that has been solarized tends to retain more microbial life than if the bed had been occultated; that is, as long as you have devised a good plan for putting the clear plastic in place and taking it off in a timely way.

Studies have shown that extended periods of solarization can have detrimental effects on microbial populations and biomass at 14 days, so getting the plastic off as quickly as possible would be advised. O'Hara notes that he has noticed no significant damage in his fast solarizations. "After almost ten years of observation," writes O'Hara, "it does not appear that soil biology is significantly damaged by these quick exposures to high temperatures."

Start by mowing or crimping the spent crop to get rid of residue that might hold the plastic up off the ground surface or poke through the plastic. If there are any large holes in the plastic, consider using some UV-treated tape to repair them; otherwise, the holes could allow heat to escape, reducing the overall efficacy of the process. Solarization will not be effective unless it is sunny. But one or two sunny days should provide all the heat necessary to terminate most crops (cover crops may be an exception, as discussed in the "Cover Crop Termination" section of chapter seven).

There are many ways to secure plastic, but the goal is to ensure the plastic is tight against the ground so that wind cannot easily move beneath it, and so the heat stays trapped. Sandbags and weights work well.

Maximizing the efficacy and speed of solarization will require some monitoring and practice. I recommend checking the temperature regularly (I keep a basic thermometer poked through one part of the center of the plastic that I can check every day). If it is not reaching that 125°F (50°C) mark for extended periods, it won't likely kill the residue as quickly. Once the plastic is removed, replant as quickly as possible.

CHAPTER SIX

Path Management

Permanent beds are a staple of no-till gardening, but when beds are a fixed feature, so are the walking paths in between them. Those pathways constitute a substantial area that rarely has any direct economic or caloric return for the soil or farmer. Indeed, sometimes a third or more of the garden—depending on the design layout—produces nothing, yet still requires consistent management.

That said, pathways don't have to be a burden. With the right management plan, paths contribute to a healthy garden environment, serving as moisture retainers, microbial gatherers, erosion buffers, and more.

This chapter focuses on how to decide on and implement a path management plan. The goal is not to just prevent pathways from getting out of hand but to make them work for the grower. I have tried several different pathway management strategies, and I hope that my own experience and the experience of other growers described in the next few pages—using methods from wood chips to living pathways—will help you design the right approach for your context.

Wood Chips and Other Mulches

Many microbes are not equipped with the enzymatic capabilities to break down woody materials, but saprotrophic fungi are built for it. These microbes readily recycle the lignin and cellulose—the compounds that make up wood, stems, leaves, and other stiff structures in Nature—and unlock the nutrients contained inside. Using specialized enzymes, saprotrophs deconstruct branches, sticks, and leaves, making nutrients or minerals in those materials available to other soil biology and plant roots. Not to be confused with mycorrhizal fungi that attach to plant roots, saprophytes play an equally critical role in soil building and nutrient cycling. The good news is that growers can easily encourage these organisms in the

Figure 6.1. Wood chips make a great source of food for various types of fungi that break down wood chips, sticks, and leaves into plant-available nutrients.

garden. Pathways provide that opportunity—just fill them with woody materials like leaves, bark mulches, or wood chips. Or all three.

WOOD CHIPS

To help mitigate compaction and retain soil moisture, apply wood chips to a depth of four to six inches in pathways. You probably don't want the chips to stand taller than your beds (the exception could be gardens in an arid climate where very deep chip reservoirs in pathways can provide some moisture retention by protecting the beds from wind). As discussed in chapter four, the concern that wood chips may tie up soil nitrogen applies only to a shallow depth in the underlying soil. In pathways that issue is minimal—they are more likely to be a source of nutrients than an element that will steal them.

Grower Charles Dowding has been a strong advocate for wood-chipped pathways. He finds that the crop rows closer to the pathway in his gardens tend to do better during drought years than plants in the middle of the bed.

However, wood chips may not be the most reliable mulch if you live in an area where heavy rainfalls of three inches or more at a time are a regular occurrence. This is especially true if you're growing on a slope, as I've learned from several experiences. Wood chips will readily wash out of pathways or onto the beds in heavy rain events, especially in the first month or two after application. Once the wood chips have established, the

Growing Mushrooms in the Pathways

To help keep wood chips from washing out, and to add some economic potential, some growers propagate saprophytic mushrooms such as *Stropharia rugoso-annulata*—wine cap or king stropharia—which is a large edible mushroom. The *Stropharia* mycelium expertly colonizes the wood chips while providing some economic benefit in the form of delicious mushroom fruit.

I endorse wine caps as a way to try out mushroom growing because they are easy to propagate, produce a substantial amount of biomass, and are all around tasty. In our experience, however, the aggressive fungal mat created by the wine cap mycelium is very effective through moderate rain events, but it does not quite hold up through heavy rainfalls. Also the mushroom caps themselves, albeit still delicious, turn brown and dry out fast if grown in full sun, thus becoming less marketable than the burgundy red caps that form when proper shade is provided. In terms of keeping the rain off the mushrooms and enjoying them, greenhouses are good options for wine caps, as are special wine cap beds beneath perennials. Hedgerows that include large shrubs, and the pathways beside them, are excellent places to employ this mushroom.

Figure 6.2. Wine cap mushroom mycelium in wood chips and bark mulch.

saprophytic mycelium will help hold them in place, locking the wood chips together with strings of white fungal hyphae.

One potential disadvantage of wood chip paths is that a layer of wood chips alone will not (generally) be enough to prevent all perennial weeds. Many weeds, including bindweed, bermudagrass, nutsedge, and others can make their way through even thick applications of wood chips. And once you have spread wood chips in paths, your only option for weeding is by hand. For those reasons, I suggest first putting down a thick layer of cardboard and then laying down the wood chips. Or, take a season to cultivate out all of the weeds in your pathways before applying any wood chips. Doing both assures the highest rate of long-term weed suppression.

Almost any tree species can be used as wood chips. A mixture of chips from different trees will be best for a diversity of nutrients. If you can, opt for ramial wood chips, which are chips generated from young branches of deciduous trees. These branches, generally less than three inches in diameter, contain the highest concentration of nutrients of any part of a tree. The concentration of essential minerals and nutrients—nitrogen, phosphorus, potassium, calcium and so on—intensifies as the diameter decreases.

The Université Laval in Quebec has been the leader on research about ramial wood chips, and Quebecois growers such as Jean-Martin Fortier and his partner Maude-Hélène Desroches are long-time advocates for adding ramial chips to pathways. They pile the ramial chips in the pathways, leave

Figure 6.3. Cover-cropped beds with ramial chips in the pathways at Les Jardins de la Grelinette in Quebec. The chips come from a local purveyor of ramial chips made from willow tree branches.

them for one season to break down, and then shovel them onto the beds using a rotary plow—a corkscrew-like implement on the back of a walk-behind tractor that can easily throw the chips to one side. Relocating these chips from the paths to the beds every year provides rich organic matter, as well as a nutritional boost, to the growing area. In effect, they are using their pathways to create a nutritious mulching compost for their beds. It's a great use of generally unused space.

Other carbonaceous mulches may serve better than wood chips in pathways in tropical and subtropical environments where heavy rainfalls are common or where wood chips are uncommon. Straw and hay are both options for pathways, though contamination with seeds and residual herbicide will be a concern. This concern extends beyond the pathways because, practically speaking, when you're spreading hay into a small pathway, it is nearly impossible to prevent some hay or straw from landing on the bed tops, thus potentially introducing weed seeds in the growing areas.

CARDBOARD AND PAPER

Cardboard and paper mulches are somewhat difficult to apply to established pathways because they tend to be wider than the average path. For example, the minimum width of WeedGuardPlus is 24 inches (61 cm). Paper mulches do not stay in place on their own, and they must be paired with another mulch, such as compost or wood chips, to hold them down. However, if

Figure 6.4. Wood chips delivered directly by a tree service company often include a variety of tree species and branch sizes plus leaves and bark, making it full of many potential plant nutrients.

Figure 6.5. I use a broadfork to decompact pathways where water tends to collect. Otherwise, water will eventually reach the level of the bed tops and rush over the beds, eroding the mulch material.

you're setting up a new garden, then it is possible to cover the bed tops and walking paths with cardboard or mulch paper simultaneously. In this scenario, the beds would get formed with compost and the pathway would remain as bare cardboard. I like using cardboard as an additional barrier under mulches in general where time and material allow.

Earthworms are fond of corrugated cardboard, which—because earthworms provide so many nutritional and physical benefits to the soil—makes cardboard an attractive mulch for improving soil health and texture. The reasons why earthworms like cardboard aren't known for sure, but research suggests that it may be due to the glues used in cardboard production. The glues may provide a rich protein source for the many microbes that worms eat. In my experience, even bare cardboard in pathways is effective, but it can become slick to walk on and it breaks down within just a few months. Thus, for long-term weed blockage, be sure you have a strategy in place for replacing worn-out cardboard as needed, or for keeping any cardboard in pathways covered with another carbonaceous mulch.

SAWDUST AND BARK MULCH

Sawdust and bark mulches are good options for pathways. Do not lay dyed bark mulches, because they may contain unwanted chemicals. If you can find natural bark mulches affordably, you're in luck—they are high in nutrients and saprophytic microbes flock to them. Sawdust mulch can generally be purchased for cheap or found for free at lumber mills. The small particle sizes of these materials make them compact together tightly, and so they allow less weed penetration. Due to this strong matting effect, and their somewhat water-repellent nature, they can remain in place during rain events. It is not as necessary to apply a layer four to six inches deep as it would be with loose wood chips. Two inches should be fine in most paths as long as they are cleared of weeds first. For sawdust or any fine mulches, apply them before the beginning of the season when there are not leafy greens in the garden beds that can be covered in sawdust and thus be hard to clean at harvest.

WOOL

Some growers have access to spoiled, unsalable sheep's wool. This material can make an excellent weed mat, and it contains a great deal of minerals and nutrients. According to a 2012 Sustainable Agriculture Research and Education (SARE) report from Turner Farm in Ohio, excess wool from their flock worked well as a garden mulch, and they saw substantial increases in yields in tomatoes, peppers, and eggplant. The likelihood of gathering enough wool mulch to cover growing beds on a commercial scale is low, but perhaps pathways could be a good use of wool. Wool might deserve more credit as a mulch, especially given that it's a naturally regenerated resource.

COMPOST

For most growers, compost is too expensive to employ as a path mulch, though it certainly works. We have found that compost we apply to our growing beds slowly works its way into the paths, via raking, wind, harvesting, children throwing it around, or the simple force of gravity. Jay Armour and his wife Polly of Four Winds Farms in New York have been employing a deep compost mulch for over 20 years. Jay has said much the same thing about their paths: They were never purposely mulched but have become passively mulched with compost over the years.

Plastic Mulches

Plastic mulches are a little trickier to use than other mulches, and not just for their environmental concerns. Our farm is certified organic and so we are not allowed to leave plastic on the soil surface after the growing season has ended. That said, I generally don't find mulching with plastic effective as a long-term weed blocking strategy. In my experience, when using plastic mulches on the ground—say around the garden or beside high tunnels—the mulch will block weeds for only one season. After that, weeds will come through, or grow on top of, the plastic. Pulling those weeds subsequently pulls up the plastic, and it becomes a very difficult and unenviable task. By contrast, although it will need reapplication sooner, an organic mulch will not only provide nutrients but generally be easier to hand weed or weed whack in the long run.

One exception is in a new garden setup. In this case, the use of landscape fabrics can be considered for creating pathways. Four-foot-wide (one meter) sheets of landscape fabric can cover both the beds and pathways when the beds are 30 inches (75 cm) wide and pathways 18 inches (45 cm) wide (to see an example, please refer to figure 4.7 in chapter four). You would plant through the landscape fabric in the beds, and leave the landscape fabric in place through the growing cycle, or all season. Occultation will occur in the pathways as a bonus, clearing them of weed seed. Those fabrics would then be removed and you could then apply an organic mulch to the now largely weed-free pathways.

Living Pathways

Because living plant roots confer so many benefits to soil, maintaining a stand of living plants adjacent to active crop beds is an ecologically dynamic approach to path management. Of course, there are some important considerations that must precede implementing this approach.

For starters, working with living pathways requires an adjustment of attitude on the grower's part—a willingness to relinquish some amount of control. In essence, attempting to maintain weed-free beds next to pathways full of potential weeds causes dissonance; our affection for symmetry competes against our love of ecology. It feels risky, wild—and great—all at the same time.

Successfully mitigating the inherent challenges of living pathways requires good planning and preparation. If you decide to attempt living pathways, start small—one plot, a handful of beds, nothing too overwhelming. It's easier to expand and put more pathways into perennial crops the following year than it is to remove the sod from a too-large experimental area.

The first step is not, as you might assume, choosing what types of plants to grow in the pathway, but to figure out the appropriate path width. I discuss bed and path width design in chapter two, but living pathway widths require further consideration.

Generally speaking, decide on the mower you intend to use to manage your pathways before setting the width of the pathways themselves. Many kinds of mowers are adequate for living pathway management, and numerous gas, electric, and self-propelled mowers on the market are a good fit for the task. The mower blades should be wider than the wheel base (so the wheels will not run over plants as you mow paths). The mower should not

Figure 6.6. Planting a living pathway of white clover and ryegrass between okra beds gave us our best okra harvest ever.

Alternatives to Mowing Pathways

A weed whacker can be used to manage living paths in a backyard garden or microfarm, but the task can be quite messy and labor intensive. It is also possible to make living pathways wide enough that they can be managed with livestock and portable electric fencing, though for most vegetable production I do not recommend this approach. The attendant food safety concerns when managing livestock near crops are too great, and positioning ruminants right next to tender plants far too risky. For orchards, however, animals are an excellent option, but you'll need to develop some animal husbandry experience before utilizing animals in your orchards or around your gardens.

throw clippings off to the side. Flail or mulching mowers are the ideal option because they direct the clippings back down onto the sod. Ideally, you will not have to buy a separate mower for pathway management. The mower you already use in your yard or garden should suffice. But if you do buy a mower for this job, choose one built to last, because it will see no lack of work in the course of a growing season.

The width of your mower is the determining factor for the width of your pathways. There are several 18-inch (45 cm) mowers—both electric and gas powered—on the market, though a 21-inch (50 cm) deck is more common. Sixteen-inch (40 cm) mowers exist, but they tend to be less robust or nonmotorized. Designate your path width based on the mower—they should be the same width.

SOD VERSUS MIXTURES

The next choice is whether or not to maintain the existing sod in pathways. If you intend to maintain the sod that is already in place, know that some grasses (bermudagrass for instance) that spread via rhizomes can be highly invasive. Be sure to identify the grasses in your pathways. If you don't know the identity of a grass, try referring to online resources, including gardening message boards and groups. Ask another farmer for help if need be. If you do decide to retain the existing sod, I recommend broadcasting clovers and ryegrass and other beneficial perennials into that sod so as to fill in any gaps in the grass, maximizing that area's photosynthesizing potential.

If you decide to get rid of the sod, then you really have two options. The first is to mow the sod very low, cover it with compost, and sow what you would like to be there. The second is to simply tarp the entire plot until all the perennial grasses have died and then establish the plants you would like. In a situation where you have wood-chipped pathways, but would like to try living pathways, simply shovel mulch onto the beds and sow the pathways with your desired crops.

If your pathways have no existing sod and you're starting with a blank canvas, consider planting a mix of multiple species. A mixture of clovers and perennial ryegrass is a good place to start. Creeping thyme, chamomile, and other low-growing perennials could be mixed in as well (not to mention they provide an aromatic treat when you walk on them). I think having a mixture there of perennial legumes such as white and red clover is a good choice to start. Once you've sown your pathways, though, it's important to understand that, over time, Nature will ultimately determine what grows in your pathways—unless you wish to do extensive hand weeding, weeds are bound to appear in living pathways. This can result in a change in the species mix at different times of the year. We've found that clovers do well in the spring and fall, whereas grasses seem to take over in the summer. As long as we mow before the grasses form seed heads, though, the grasses do not become problematic.

Jennie Love of Love 'n Fresh Flowers and host of the *No-Till Flowers* podcast manages three acres of flowers in a no-till system in Philadelphia.

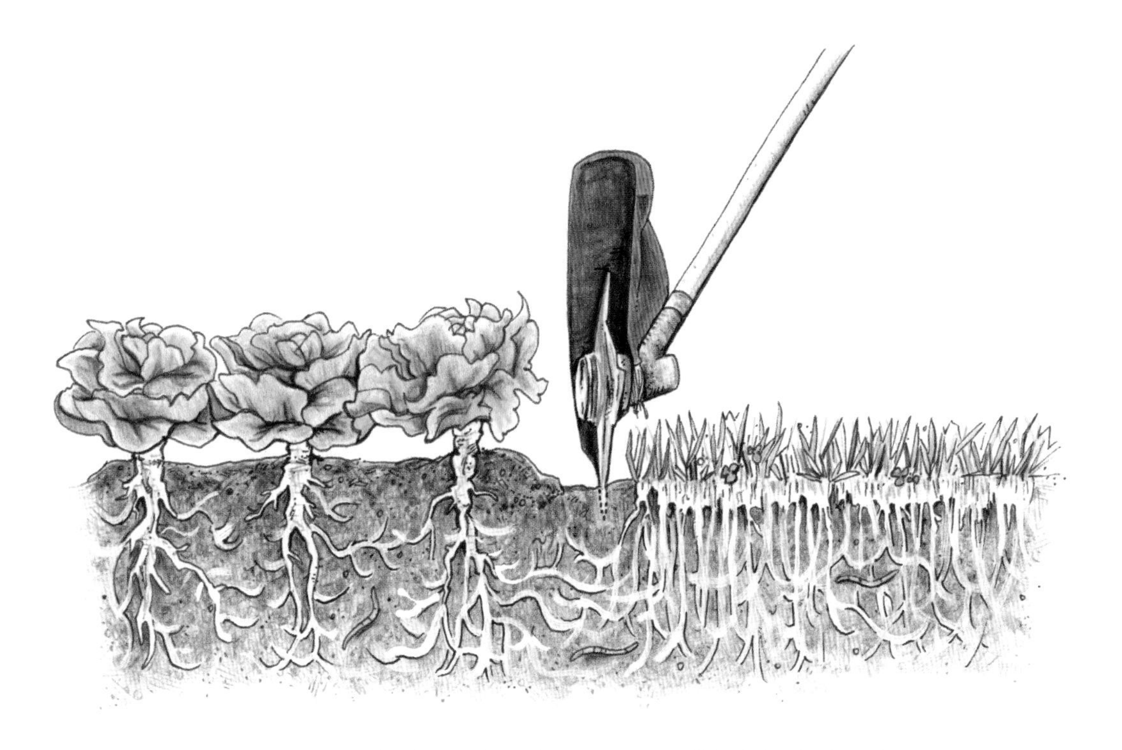

Figure 6.7. Run an edger along the margins of beds to prevent weed creep. The blade does not need to go more than a couple inches deep. That way, many of the beneficial belowground connections will stay intact.

Jennie maintains living pathways in many of her beds. She emphasizes the importance of managing them consistently, which includes mowing them every week during the growing season to reduce any potential for creep.

The margin between the beds and pathways can be managed with an edger, or alternatively with a weed whacker turned vertically (if you're not familiar with this landscaping technique, it's easy to find videos online that demonstrate it). This shallow edging simply cuts off many of the shallow rhizomes that might creep into the beds, but it doesn't disturb the microbial activity that helps to funnel nutrients from the pathways into the beds, as illustrated in figure 6.7. The majority of those nutrients and microbes are situated around four inches deep—that is, around the rhizosphere, and well below the zone affected by edging. As long as the edging is shallower than four inches deep, there will be limited effect on nutrient availability but a reasonable effect on preventing pathway plants from invading the beds.

Living pathways do not need watering, but the beds beside living pathways will need consistent irrigation, even in wetter climates. Especially in the first few weeks after transplant or emergence of crops in the beds, the pathways seem to dominate the moisture supply, and we've seen crop plants suffer during a drought before the pathway plants do.

STRIP TILLAGE

The most substantial field studies on living pathways focus on strip tillage systems in which the paths are left as living ground cover but the beds are tilled every year. These studies, often paired with conventional, chemical treatments, regularly show increases in crop yield as well as fuel and energy savings. As discussed previously in this book, Helen Atthowe and the late Carl Rosato of Woodleaf Farm have done some great work in the area of organic strip tillage. They found that a number of beneficial parasitic insects used a strip of tall plants in the pathways as habitat. It's important, though, that the seed heads of pathway plants are not allowed to extend close enough to the beds to create weed issues.

Living pathways are most commonly seen in strip tillage scenarios, where the pathways are left to sod or are planted to a specific perennial but the beds are turned over every year through mechanical tillage. The practice of strip tillage is gaining attention from large-scale agriculture for its ability to conserve soil, build organic matter, and reduce runoff. A new term, biostrip till, is beginning to crop up in larger-scale production. Broadly speaking, it describes a system of allowing pathways to be in sod while the beds are sown with winter-kill cover crops (in lieu of a compost mulch, for instance, or tillage). So the beds are ready to plant in the spring without additional tillage, creating a sort of large-scale strip no-tillage.

Mulch-in-Place

One ecological way to manage pathways is through a mulch-in-place system involving cover crops. Such a system provides some benefits for market crops that will manifest the following spring after the cover crop is sown. These benefits include nitrogen fixation or weed suppression through allelopathy (when chemicals secreted in the exudates of one plant inhibit the growth of others), and the cover crops themselves also provide a small amount of mulch for the paths. I discuss winter-kill and termination of cover crops in chapter seven, and these practices are much the same for both garden beds and pathways. In essence, a given area is sown with a dense cover crop that can winter-kill or be crimped; and the resulting mulch will suppress weeds.

Figure 6.8. These field peas sown between fall carrots will be killed by exposure to a few freezes over the winter and will serve as a light mulch the following spring. As legumes, their presence generates plant-available nitrogen for use by next year's market crop.

Cover crops that do not winter-kill can be sown in pathways and then tarped or solarized to terminate them (this practice is discussed in more detail in chapter seven). Though it differs slightly by region, winter-hardy cover crops for this use include Austrian winter peas, cereal rye, vetch, crimson clover, and wheat, among others. In some areas, oats and field peas may be winter hardy. In other areas, hardy crops like crimson clover won't be—check with your local extension agent or nearby farmers for guidance. For a summer crop, sunn hemp, sorghum sudangrass, and *Sesbania* species can be used. Sowing a mixture guarantees something will work, and it increases soil diversity. I dive into more details about all the different cover crops in chapter seven, and also how to terminate each.

Any path can be sown with cover crops as long as the pathways are already clean of weeds and loose enough to seed. How you sow the cover crop will be partly based on what tools you have and what kinds of seed you are using. If the soil in your paths is not compacted, most seeders will do the trick. If the soil is hard and overly compacted, you will need to carefully broadcast the seed by hand and then cover it with peat moss, a light straw mulch, or compost. Water that in well if possible to achieve rapid germination and to prevent birds or other pests from finding the seeds.

If the paths are not too weedy, another straightforward method is to broadcast the seed in the pathway then lightly cultivate the area. Cultivating will eliminate weed competition and lightly bury the cover crop seed in the process. I do recommend lightly covering the seeds with a little compost and walking back over the area to increase soil-seed contact. Simply broadcasting seeds without covering or pressing them into the soil will result in very limited germination.

Note that sowing a cover crop next to beds you intend to use through the fall is an effective way to establish a winter cover, but the presence of the cover crops may be annoying at harvesttime. Moreover, the cover crop can grow large enough to begin to block out sunlight or block airflow to crops. In general, the goal is to time the sowing of cover crops so that they gain sufficient aboveground biomass before the fall but not to go to seed. Err on the side of caution. See the "Critical Period of Competition" section on page 181 as a guide to timing the sowing of cover crops so as not to cause issues for the intended cash crop.

The mulch that results from using cover crops in this way will last well into spring, but likely not through the summer. It will provide many of the benefits of a permanent living pathway without as much competition. If nothing else, cover crops make a much more ecologically beneficial "placeholder" over the winter than plastic, because the soil will be filled with living plant roots.

No Mulch in Pathways

Whether you spread straw, sow a cover crop, or allow compost mulch to passively migrate into pathways, ideally you are always working toward maintaining a pathway mulch of some form. There are too many benefits involved not to make this a priority.

However, getting weeds under control before you mulch your pathways will serve you well for years to come in terms of labor savings. Especially in warmer, wetter climates such as the Southern United States, where the "weed season" lasts for several months, keeping weeds in check is almost a prerequisite to path mulching. In drier climates, a year of path cultivation may not be necessary. In our warm, rainy climate, we've found that once a grass or perennial weed makes its way through wood chips or other mulch, the only option for managing the weeds is to pull them by hand. Obviously, that is neither an efficient nor desirable way to spend our time.

Managing pathways without mulch takes some specific tools and disciplined commitment. Without the mulch to assist in blocking the weeds, cultivations must be regular and preemptive—battling sod in the pathways is demoralizing and time consuming.

Figure 6.9. Whether our paths are mulched or not, our most common pathway weeds are dandelion and grasses, which we keep in check with the assistance of a stirrup hoe.

When not using a mulch, set aside time to cultivate the pathways every week during the growing season using a stirrup hoe, a wheel hoe, or cultivating shoes on a tractor. Cut off all weeds as low as is necessary to terminate them. A collinear hoe is not an effective tool for this task; the hard soil in pathways can easily break the blade. Ideally, after a cultivation, rake the plant matter together and pull it out of the pathways so it does not reestablish. That step is less necessary on sunny, dry, hot days. Keep the blades of any hoes for this job sharp so as to speed up the process and make the tools more effective. Do not step on areas you have cultivated; doing so can effectively replant the weeds.

The first season of unmulched pathways will be the hardest and most labor intensive to manage. Once the weeds are in check it is possible to start exploring more permanent mulching options for all that space between your beds.

PART THREE

Keep It Planted as Much as Possible

CHAPTER SEVEN

Fertility Management

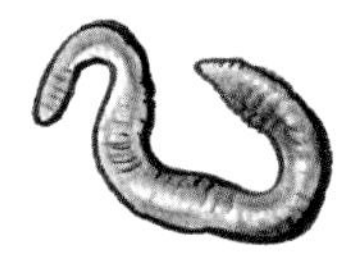

Managing soil fertility is often thought of in terms of physical additions to the soil—composts, minerals, fertilizers, or any other amendments. Check your soil test results and add whatever the report says is missing or out of balance. But simply adding amendments to a soil will not make it fertile. What makes a soil fertile is how well prepared it is to support healthy plant growth.

Certainly, there are times and situations in which adding a particular amendment helps make the soil more prepared to grow crops. In this chapter I expand the concept of soil fertility management to encompass more than just amendments. Compost or any other amendment on its own is simply not enough. A good fertility program must also take into consideration all the primary soil conditions that support plant life and photosynthetic activity—consistent moisture, a constant supply of organic matter, good soil structure, permeability, thriving microbiology, and living plant roots. If we do not tend to those six components of healthy soil, nothing we add will make our plants thrive.

Measuring and Managing Fertility

The six factors in a good fertility program work in concert, but it's important to deconstruct them and describe how to evaluate and improve upon each. Keep in mind there is no single measurement to tell you how healthy or fertile a soil is, because there is no single factor. Each constituent part of soil health must be considered, evaluated, and managed appropriately. It's a process and it can take time to develop. As such, I describe here a long-term approach to soil fertility not based solely on what crops need to succeed, but what soil requires. And this goes well beyond simply adding amendments.

WATER

Almost all the elegant, intricate chemistry of life is mediated by water. Water transports nutrients and microbes. It is essential to photosynthesis and is bonded with various nutrients to feed our plants. So intrinsic to life is water that astrophysicists look for water as the primary sign of possible life on other planets. Without water, life cannot thrive—soil life not excluded.

It starts with the quantity of water. Too little moisture and soil life perishes. Too much water and anaerobic (often pathogenic) microbes take over. Inconsistent moisture can also have grave effects on living soils because after the soil dries out, soil microbial populations don't immediately bounce back. Clay soils, compacted soils, or soils low in organic matter tend to suffer from poor drainage. If you see standing water during rains, or if the soil is muddy to the touch or smells like rotten eggs a few days after a rainfall, that's an indication of poor drainage.

Indeed, soil should feel moist and smell neutral or sweet. Water must be able to penetrate during a heavy rain or at least shortly after. If not, issues of poor drainage have to be addressed by breaking up compaction, keeping the soil filled with living roots, and inoculating the soil with compost teas and extracts. Raised beds and deep mulches also help. In extreme examples where standing water is common, drainage can be installed, but this is a last resort. Instead, try to pick growing areas that do not have chronic drainage issues (those waterlogged areas are where you should dig ponds).

The quality of the water used in the garden is likewise paramount. Municipal water usually contains antimicrobial chemicals intended to protect consumers from, for example, harmful bacterial buildup in water pipes. But like most antimicrobials, the chlorine and chloramine added to municipal water supplies are indiscriminate and can kill beneficial microbes as well as pathogens. So this sort of water does not support soil life (or, for that matter, gut flora). That said, well and spring waters are not inherently void of contaminants. Nitrates and pathogenic bacteria such as E. coli are among other possible contaminants in groundwater. If your water source is at all suspect, it is worth paying for yearly water analyses and adding a high-volume filter to your irrigation system if necessary.

A market garden soil also cannot be fertile unless healthy water is applied evenly and consistently. When soil lacks moisture, soil respiration and nutrient cycling can slow down, and rewetting dry soil causes a burst of biological activity that consumes soil carbon.[1] An effective irrigation system is key, and if water is in short supply or expensive to access in your area, then an efficient system is also paramount. If you do not have experience with irrigation, the best recommendation for designing the right irrigation system is one I received from Jean-Martin Fortier: Ask an irrigation

Dry Farming

Can a soil be dry farmed? Absolutely—we dry-farm several plots on our farm every season, including our sweet potatoes, which we do not irrigate. When dry farming is the goal (or a necessity), give crops more space—much of the intensive planting we do in our irrigated gardens would not work in our dry-farmed plots. If crops are not competing with one another for moisture, the soil will be less likely to dry out. Also consider the soil type and organic matter content—sandier soils, or soils low in organic matter, tend to dry out more rapidly than denser, richer soils.

Do not use raised beds for dry farming in any region where moderate drought is a possibility. Although rainfall is fairly consistent in our Kentucky location, we can experience droughts. Some growers do not have a choice—certainly water can be extremely scarce in many regions of the world. But dry farming should not be a dogma. Some type of backup water supply should always be available for emergencies and for "watering in" crops at transplanting or seeding. Crops that begin their growth in dry soil will be at a disadvantage from the start. The plants and the soil microbes need that moisture to get started. Deep mulches and all other practices described in this book can dramatically improve soil moisture retention.

Figure 7.1. A dry-farmed sweet potato patch with meal corn interplanted.

supplier for help. There is no better way to design an irrigation system specific to your farm and climate than to have an expert, preferably a local one, assist you with it (and they will generally help you simply in exchange for the sale of irrigation equipment). Also consult with other farmers and visit farms in the area that use a similar water source.

SOIL ORGANIC MATTER

Soil organic matter is loaded with nutrients for plants. It retains water in a drought and, almost paradoxically, redistributes water after a thorough drenching. Soil organic matter acts as food and habitat for microorganisms. The soil's ability to hold on to nutrients is heavily reliant on its organic matter content, and it is said that soil organic matter makes up 90 percent of soil function. In effect, even though soil organic matter is largely a bunch of dead organic materials, a truly living soil must contain a fair amount of it.

Figure 7.2. Carbon compounds are dark in color. The soil core in this soil probe gets lighter from top to bottom, indicating a lower amount of organic matter content deeper in the soil.

Soil organic matter is among the more helpful analytics that can be found in a soil analysis. The soil organic matter percentage should increase or at least stay the same over time. An improved percentage of organic matter in the soil is a great indication that your fertility programs and soil management activities are working, because it means roots and microbes are getting down lower in the soil and digging up deeper nutrients and minerals.

Between 5 and 10 percent of soil organic matter is considered to be a healthy percentage. If your soils are in this range, you don't need to take any dramatic steps. Just follow good soil practices to maintain that percentage as described throughout this section. More than 10 percent organic matter is considered unnecessarily high, and is most likely the result of excessive mulching. For the most part, there are no significant benefits that I can find in the scientific literature to having very high organic matter in most soils. There can even be some downsides when organic matter exceeds 12 percent, such as difficulties with moisture retention, and potential for leaching nitrogen and phosphorus. The pore space between particles can be so large that microbes cannot create the aggregates that help retain water (this is easily observed in mulchy composts and is often why growers complain of poor germination when they seed crops into compost—the lack of moisture retention allows seeds to dry out or prevents seedlings from rooting well). If you are seeing negative effects of high soil organic matter, just pull back on adding compost or mulching materials. The problems will clear up quickly.

Low organic matter can be partially addressed when setting up a new garden (described in chapter two) by injecting a robust amount of compost and activated biochar through a one-time tillage. Otherwise, building organic matter is a long-term process, but an important one.

Keeping living plant roots present in the soil is the most efficient means of generating soil organic matter—cover crop roots in particular. Any plant roots—crops or cover crops—that can be left in the soil after a crop is terminated will be a significant contributor to soil organic matter. Interplanting (described in chapter eight), is a great way to not only ensure a diversity of plant roots but to keep the soil full of plants at all times. Mulches contribute soil organic matter to a smaller degree, as soil organisms shred the mulches from below, burying and mixing that material into the native soil. Soil organisms themselves, living and dying, are another notable component of organic matter.

SOIL STRUCTURE

No aspect of soil fertility operates in a vacuum—all constituent parts of living soil depend on one another. Soil structure is a prime example. Soil structure is created and maintained by living plant roots and soil organisms. Worms and roots push their way through the soil, creating tunnels. Moles and voles burrow around, opening up pockets for air and water while depositing their manure throughout. Microbes aggregate the soil into crumbs and clumps of different sizes, which renders the soil friable. Soil structure is the result of the sum total of all these actions, and that structure is vital because it allows organisms, water, oxygen, nitrogen, and carbon dioxide to move more freely through the soil.

Especially in the first few years of a garden, monitoring soil structure can be helpful. Simply plunge a probe or shovel into the soil about 8 to 12 inches (20 to 30 cm) deep to collect a sample. Gently break apart the lowest section of the soil core from the probe (or the soil slice from the shovel blade). Notice whether roots are penetrating and pore spaces are

Figure 7.3. Earthworms consume microorganisms and organic matter, and that is why their presence can be an excellent indicator of living soil.

Figure 7.4. When we avoid tillage, our soils can retain their healthy structure. In this soil sample from my no-till beds, the pores that allow water and soil organisms to move through the soil are easy to see.

developing that deeply. Examine portions of the core or slice starting from the lowest section and moving to the soil surface. The sample should become increasingly dark, porous, and crumbly.

If it is difficult to even push the probe or shovel blade into the soil to collect a sample, it's a sign that the soil has compaction layers. Compaction limits soil permeability and thus the development of soil structure. Your first job will be to address the compaction—through broadforking or whatever method makes the most sense given the depth of the compacted layer. (I discussed methods for remedying compaction in chapter two.) After that treatment, it's critical to maintain living roots in the soil as much as possible. Plant roots should be kept in the soil at all times. I also like to soak transplants or drench the soil with extracts of compost teas at planting to encourage microbial populations (see the "Compost Teas and Extracts" section on page 145). Doing this helps to create and improve soil structure.

SOIL PERMEABILITY

Ground-up rocks are the predominant ingredients of soil. The size of the rock particles determines soil type: Clay particles are extremely small, silt particles are midsized, and sand particles are large (in comparison to silt and clay). Most soils consist of a range of particle sizes—the most coveted mixture is loam that, for its part, is 40 percent sand, 40 percent silt, and 20 percent clay.

The size of the majority of particles makes an enormous difference in how a soil performs. If it is too dense, as clay soils often are, drainage and root penetration may be poor. In light, sandy soils, nutrients and moisture are hard to retain. The most accurate way to know what kind of soil you have is to consult your local extension agent. Each county is likely to have its own data set for soil types by area. In the United States, the Natural Resources Conservation Service (NRCS) maintains soil maps and data online at the Web Soil Survey website (other countries may have their own organization that does this work). At this website, you can learn precisely what type or types of soil you are working with on the land where you farm or garden.

Can you change the physical nature of the soil? Yes and no. You can't change the soil particle mixture you have—clay, silt, or sand—without hauling in expensive truckloads of topsoil, which is not generally a viable approach and comes with its own issues of sourcing topsoil free of contaminants. But you can make any soil perform and feel similar to loam, at least over time.

The extreme soils—heavy clay and sand—are the most challenging to improve. When working with heavy clay soils, consider blending in some amount of organic materials such as a rich nutritional compost, complemented immediately with a long season of cover crop. There is no simple recommendation for how much compost to apply because of variation in

Natural Farming and IMO

Natural farming is a term that refers to a specific cluster of practices that are widely used from Japan to Korea and throughout the Asian world. In large part, thanks to the work of Han-Kyu Cho of South Korea, the most well-known form of natural farming in the United States is generally referred to as Korean Natural Farming (KNF).

Along with his family, Han-Kyu Cho has done extensive work in creating, and adapting from other natural farmers, simple methods for capturing indigenous microorganisms and unlocking all the macro- and micronutrients that plants and soil biology need to thrive from local resources.

Indigenous microorganism (IMO) collection is the centerpiece of Korean Natural Farming. Indigenous microbes are fungi, bacteria, archaea, microarthropods, and other microscopic biology already adapted to thrive in one's area. Rohini Reddy describes this process in detail in the book *Cho's Global Natural Farming*. The "Simple Inoculating Compost Recipe" on page 61 utilizes some of those strategies, with some key differences.

In the Korean Natural Farming method, indigenous microorganisms are cultured on steamed rice. The process is to steam the rice, place it in a wooden box that is then covered with cloth, and

Figure 7.5. Our IMO collection box.

set the box in the forest. I like to put the box inside a live animal trap to protect it from predators that are attracted by the rice. I also cover the box with a tarp for several days until visible fungal hyphae form on the rice. This rice is then mixed with brown sugar to arrest the microbial growth and make the fungi sporulate through osmotic pressure (the sudden absorption of moisture). This mixture has now become stabilized for storage by the osmotic pressure. Next, small amounts of IMO are mixed—often along with specific other KNF amendments such as fermented plant juice (FPJ), and oriental herbal nutrient (OHN)—into a carbonaceous material such as rice bran or wheat bran. This mixture is then allowed to compost through a similar process as described in the "Simple Inoculating Compost Recipe" section.

IMO collection is not the only element of natural farming worth noting. There are recipes throughout the KNF literature for practically any type of nutrient deficiency. These recipes include water soluble calcium made from egg shells and rice vinegar; lactic acid bacteria cultured from milk and the rinse water of rice; nitrogen from fermented fish; and many others. For further information on these and other KNF methods refer to *Cho's Global Natural Farming* by Reddy, the work of Chris Trump at www.christrump.com, and other references included in the bibliography on page 265.

Figure 7.6. This fungal clump formed as part of the IMO collection process.

density, even within clay soils. Start by tilling in at least a couple of inches of compost. Applying gypsum is also sometimes recommended to help break up clay soils. Always consult an agronomist before adding mineral amendments, because adding minerals may put your soil out of balance in other ways.

As mentioned previously, soil organic matter can help the soil both redistribute and retain moisture. For sandy soils, which shed water rapidly, building organic matter and adding organic materials is just as important as it is for clay soils. Blending in some well-made nutritional composts can likewise help to improve the structure of these extreme ends of the soil spectrum.

Checking soil texture is the best way to monitor the effects of your soil improvement practices. Clay soils tend to be dense and slick between the fingers. Sandy soil feels gritty, like beach sand does. No matter what type of soil you start with, as your soil improves it should begin to feel more crumbly and loose to the touch deeper and deeper down into the soil.

SOIL BIOLOGY

Soil biology is undeniably a complicated subject, but as growers we should understand the basic soil food web, a term popularized by microbiologist Dr. Elaine Ingham that refers to a network of essential underground food chains on which plants rely.

Figure 7.7. Mushrooms fruiting in our summer carrot patch—these fruits of fungi being a common sight on our farm.

In the soil food web, microbes such as fungi and bacteria use various enzymes to "harvest" nutrients from soil particles and organic matter, which then become available to the plants when the microbes are consumed or die. Another mechanism for nutrient exchange may be the rhizophagy cycle, which involves microbes actually entering *into* plant roots. This relatively new discovery is fascinating, but it is not yet fully understood. Nematodes, protozoa, and microarthropods are the main predators of those microbes. Those larger organisms are themselves consumed by other larger organisms such as earthworms, and so on up the predatory ladder. The waste products generated by all of these consumers make nutrients available for plants and increase soil organic matter.

All of these microscopic organisms can and should be added to soils through biologically active teas, extracts, and direct applications of inoculating compost. As I describe in the "Designing a Fertility Program" section on page 142, microbes are among the most important pieces of such an effort.

Microbial soil tests are available that are worth exploring. For example, the Haney Test examines soil organic matter and pH, the quantity of nutrients available to soil microbes, as well as the carbon to nitrogen ratio of the soil, soil respiration rates, and the microbial biomass. Then there are at-home, do-it-yourself microbial tests. These include the microBIOMETER soil test, which measures the fungal to bacteria ratio of a soil. Simple amateur microscopy is an option as well. This means you'll need to learn how to evaluate the microbial content of your soil through use of a microscope (there are many helpful video tutorials online about microscopy, or you can take an online course at Dr. Elaine's Soil Food Web School). I generally recommend using the Base Cation Saturation Ratio (BCSR) test to get your gardens started, and then employing some form of biological evaluation to monitor the success of your practices once the garden is up and running.

LIVING PLANT ROOTS

As I've stated before, productive photosynthesis is both the goal of living soil and the driver of it—soil needs plants and plants need soil.

Soil microbes feed on the carbon that drips from living plant roots; it is their energy source. Plant roots tunnel through the soil, making it more pliable while also leaving passageways for organisms, water, and oxygen after the roots die. In short, living plant roots build structure, sequester carbon, feed microbes, and gather nutrients for future plants—all are reasons why we leave the roots in the soil when a crop is taken out.

But the diversity of those plant roots and their durability is important, as well. Growing a diversity of plants encourages a diversity of microbial life and nutrients. As for duration, keeping growing roots in the soil for long periods of time is a long-proven way to improve garden soil. This is why so many indigenous cultures embraced fallowing fields, thus taking them out of production. When those fields came back into production, they performed better as a result of the years they spent completely undisturbed and covered by photosynthesizing plants. Of course, given the small land bases that many market gardeners are working with, fallowing is not always an option. Cover crops are an excellent alternative where space allows—indeed, cover crops can be the entirety of your fertility and mulching programs when done right. But again, it takes space. Let's explore how to combine all of these elements of soil fertility in a soil fertility program that is designed to make sense for your context.

Designing a Fertility Program

Weaving all the factors that contribute to soil fertility into a fertility program is not as complicated as it may seem. When a bed comes out of production, a good soil fertility program addresses any compaction, adds requisite amendments, makes sure the water is in place, saturates the plant starts with beneficial microbes, and makes sure there is some sort of mulch layer. Then the crop is planted and the bed is watered well to give it the best start possible—to make sure it's as fertile as possible. To illustrate how to turn this list into a real-life program, I will describe the fertility program Hannah and I have developed on our farm. The precise details may or may not be fully applicable to every farm in all contexts, but it is a working model to stimulate your thinking about how to devise your own program and put it into practice.

SUPPLYING WATER

Filtered water is consistently applied to the soil through overhead irrigation or drip irrigation, depending on the crops. We prefer overhead where possible because it keeps the entire soil surface moist, rendering the full garden habitable for microbes, rather than just the bed tops. (In more arid climates where water is less available, do the best you can—use mulches to both retain and spread the moisture.)

All of our beds remain covered with mulches to keep the soil in place and its structure intact—protected from the sun and wind and heavy rain. The mulches help feed the microbes while retaining moisture. They are not the main driver of soil organic matter, but they do contribute.

PROTECTING STRUCTURE AND PERMEABILITY

We check every bed before planting for compaction and address compaction whenever we come across it to make sure oxygen, plant roots, soil fauna, and water can move through easily. This sometimes involves a light broadforking in the beginning of the season, or the use of tillage radishes where time and space allow. Many of our garden beds are loose enough at this point to not require any decompaction—a good fertility program should get you to that point.

As described previously, we leave crop plant roots in the soil wherever possible to help build soil structure and soil organic matter. For a time these contribute some residual exudates to the soil before becoming microbial food as they decompose.

PROPAGATING MICROBES

We propagate microbes through specially made inoculating composts (recipe on page 145) that we apply to the soil in a variety of ways. We use compost extracts as soil drenches or for soaking transplant soil blocks before they are set out in the garden. We also pour the extracts into planting holes for large transplants (such as tomatoes) before setting the transplant inside. All of this helps put the microbes precisely in the place where plants utilize them most—the root zone. Compost teas made from well-made inoculating composts are often mixed into our extracts or used as foliar sprays throughout the season.

SUPPORTING HEALTHY GROWTH

We never add any amendments when preparing beds for planting if the new crop is following a cover crop (see "Using Cover Crops for Fertility" on page 148). Anytime a cash crop is not following a cover crop in our gardens, however, we add three pounds (1.5 kg) of alfalfa meal, three pounds (1.5 kg) of feather meal, and one-eighth pound (60 g) each of kelp and humic acid per 250 sq feet of bed space before planting. The two types of meals provide protein that the microbes turn into a slow-release nitrogen for the plants. In fact, I very much enjoy adding grass meals and bean meals to the soil, because they stimulate microbiology and add a range of nutrients that are rapidly converted by microbes into plant-available forms. When growing an advance cover crop is not possible, applying organic meals such as these provide some of those same benefits. Feather meal is a slightly slower-release form of nitrogen. The kelp provides all the essential trace elements as well as plant growth hormones. And the humic acid is a biostimulant that, as the term suggests, stimulates the microbes in the production of the special enzymes they use to harvest minerals from soil particles. Fertilizing composts such as composted chicken manure could also be added here in place of the nitrogenous amendments. Inoculating composts and biochar are other excellent additions before another crop is put into the beds.

What about Mineral Balancing?

Many agronomists believe we should be balancing our soils with curated applications of minerals. In theory, having access to all the right minerals would help plants protect themselves from diseases and pests, reduce weed pressure in soils, and increase the nutrient density and flavor of crops. Though anecdotal evidence abounds, my reading of the available research on soil balancing has not yet convinced me. This approach to agronomy has not been proven to the level that persuades me to recommend it as an approach to fertility management. Much of the research around mineral balancing has concluded little or no effect on yield or weed populations, and the cost of amendments per acre is high. Many soil minerals are mined, and there are environmental and labor concerns related to mining. As for the nutritional value of the crops, there are organizations that are researching nutrient density when mineral balancing is applied versus other growing methods. So I want to be clear that I am not suggesting that mineral balancing is the wrong approach, but at this point in time I feel that the efficacy of using mineral balancing is not established enough for me to fully endorse it. However, if the mineral balancing approach appeals to you, I recommend running some trials first to see if balancing minerals improves your production. Certainly, in order for plants to fully photosynthesize, all 17 macro- and micronutrients must at least be present in the soil. In severely degraded soils or where certain nutrients are greatly underrepresented, some amount of balancing may be in order. No dogma—do what makes the most sense for you in your context.

Our amendment list changes as we observe our crops. We are always striving to bring in fewer and fewer outside amendments, or to create more of our own. We have had promising anecdotal success using Korean Natural Farming techniques to gather nutrients from our farm (as opposed to importing them). I also see potential in techniques such as Bokashi composting, which is essentially a process of fermenting kitchen scraps, which helps to make the nutrients in those scraps available for plants.

Overall, by following a fertility program like ours, you should see microbial populations and soil organic matter increase and the need for amendments slowly subside. We continually observe plant performance and study our soil test results to see what nutrients our soils or plants may be deficient in or getting too much of. Is the organic matter going up? Are we missing any key nutrients? Soil tests can yield inconsistent results, but these tests are still insightful and provide an interesting snapshot of what's going on in the soil (and how your methods are or are not improving it).

Compost Teas and Extracts

Compost teas and extracts are liquefied versions of compost that enable growers to disperse beneficial microbes over large areas. Of all the amendments we add to the soil at Rough Draft Farmstead, microbes are the most important. Any time spent encouraging them is never wasted. Teas and extracts can be applied to the soil surface or onto plant surfaces and/or directly injected into the earth with the help of root injectors. The most common use for compost teas and extracts on our farm is to soak the root balls of transplants before planting.

Compost tea is compost that is bubbled with water and a microbial stimulant (we use a pinch of humic acid and a teaspoon of fish hydrolysate) for 24 to 36 hours. By contrast, an extract is made by letting compost sit passively in water for up to 12 hours with no bubbling to introduce oxygen. There are benefits to both methods, though perhaps more risks come with compost teas, because pathogenic microbes may more easily proliferate if

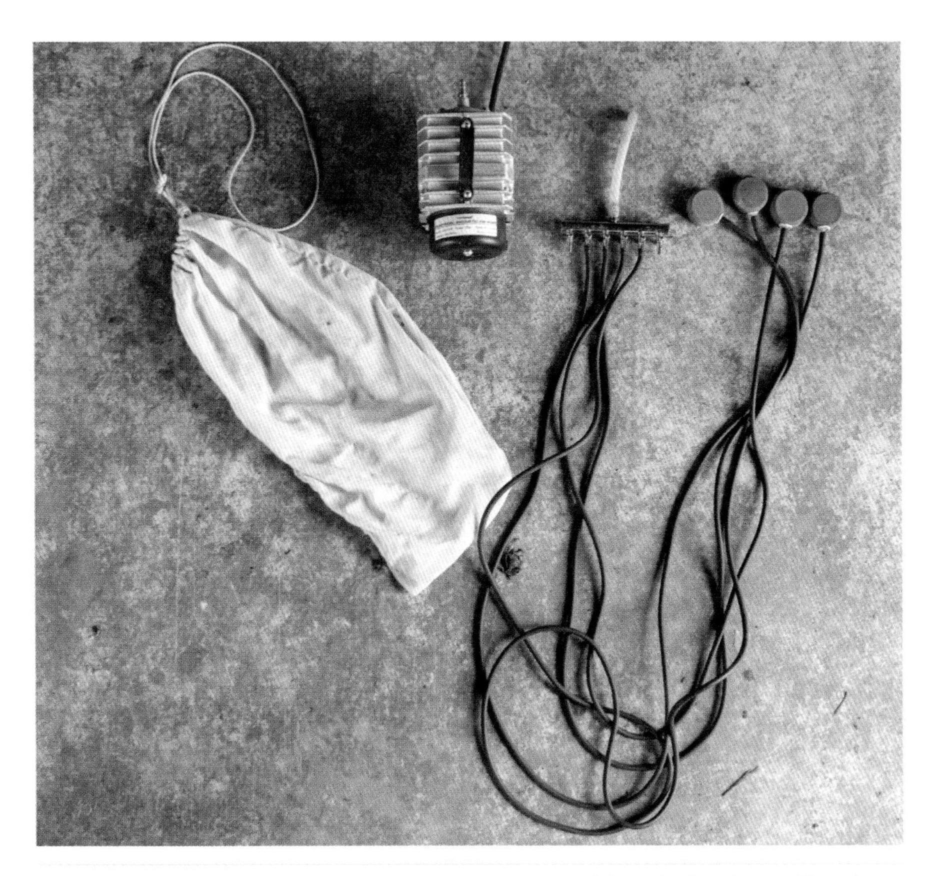

Figure 7.8. Basic compost tea or extract setup with a 400-micron filter bag, Vivosun Air Pump (950 gallons per hour capacity), aquarium tubing, and air stones. This can produce up to 35 gallons of tea at a time.

there is too much microbial food such as fish hydrolysate or molasses (a common compost tea addition). It's also possible that uneven air distribution can create anaerobic pockets in the corners of the container. Some beneficial fungi may also react negatively to the agitation, and this is why some researchers find more bacteria and yeast in compost teas but more fungi in extracts.[2] Because extracts can be made in as little as 15 minutes if need be, they come in extremely handy when it's planting time and you want to add a microbial boost on the spot. Simply add some compost to a bucket with a little humic acid, stir the contents lightly to agitate them, and then soak your transplants in the resulting liquid or pour it into the holes. Make sure to use only the best inoculating composts you have available to make extracts and teas. Where time allows I like using a mixture of compost tea and extract to ensure the greatest diversity of microbes.

Note that on our farm, even though we make NOP-compliant teas, we never apply them to the parts of plants that will be directly consumed, such as greens or fruits, so as to avoid any possible food safety concerns. It is not against the organic standards to apply NOP-compliant teas to fruits and greens, because these materials meet organic production requirements, but it is not a best practice. And, furthermore, we do wash our greens after harvest. The bottom line is that food safety should always be paramount for commercial growers small and large alike.

A Nitrogen Primer

Nitrogen is the nutrient people discuss most when talking about soil fertilization, and so I want to focus on nitrogen in the living soil context. The reason nitrogen gets so much attention is that it is one of the most significant macronutrients for photosynthesis, plant health, and human health. The human body is made up of roughly 3 percent nitrogen (and the same is true for many other animals). Nitrogen is the basis of amino acids, and amino acids are the basis of proteins. DNA, RNA, and chlorophyll are all highly nitrogenous compounds. Without nitrogen, there could be no life on our planet.

How do plants acquire nitrogen, and from where? The air around us is a major source—nitrogen (N2) makes up 78 percent of Earth's atmosphere. In terms of how plants acquire nitrogen, one important process is nitrogen fixation, which transforms atmospheric nitrogen into plant-available forms. Rhizobia are specific soil bacteria that can form a symbiotic relationship with various leguminous plants. As a result of this relationship, small organelles called nodules form on plant roots around the bacteria. These nodules help nitrogen-fixing bacteria to regulate oxygen—an important factor in converting atmospheric nitrogen. However, this form of nitrogen fixation is not limited to legumes. According to agronomist Dr. Christine Jones, all

Figure 7.9. This rhizosheath on beet roots is a formation in which microbes have gathered soil together, which helps regulate soil oxygen levels. This is similar to how nodules on roots of legumes help nitrogen-fixing bacteria regulate oxygen for nitrogen fixation.

green plants are capable of developing associations with nitrogen-fixing bacteria and archaea.[3] These other nitrogen-fixing microbes are simply much harder to study and, ergo, receive less attention from researchers.

Another primary way nitrogen becomes available to plants is through mineralization of organic matter. Stable organic matter in the soil is high in carbon, but it also contains approximately 6 to 8 percent nitrogen. Various microbes consume that organic matter, thus putting the nitrogen into a plant-available form. Protein-rich meals such as alfalfa, feather, peanut, and blood meal are also rich in nitrogen (among other nutrients). So in a nutshell (sometimes literally), when these meals are consumed by microbes, the nitrogen in those proteins becomes available for plants to use.

The digestion of organic matter and protein-rich meals largely results in transformation of the nitrogen they contain into ammonia (NH4+) and nitrates (NO3-). These are inorganic forms of plant-available nitrogen. (Inorganic, in this case, simply means containing no carbon.) However, plants can also metabolize organic (carbon-containing) forms of nitrogen, as well. They can absorb proteins, lipids, and amino acids through their roots. There is some evidence that, depending on the type of plant and time of year, amino acids are taken up for growth more than other forms of plant-available nitrogen. Temperature plays a role in which form of nitrogen plants use, and plants may also use different types of nitrogen at different stages of growth.[4] Therefore the goal should always be to make the conditions right for all nitrifying and mineralizing processes to take place so that plants can choose the nitrogen sources they need.

For growers, managing nitrogen in this way comes down to a few simple steps:

- Make sure the soil is well aerated and well structured so that there is plenty of air (and nitrogen) in the root zone.
- Regularly introduce and encourage a diverse microbial population to help ensure there are nitrogen fixers in the root zone.
- Keep living plant roots in the soil at all times so there is a steady flow of carbon into the soil to feed nitrogen fixing microbes.
- If you're adding supplemental nitrogen, choose a protein-rich material that can be converted to plant-available nitrogen in the soil. Do not apply synthetic forms of nitrogen because these can reduce microbial diversity, decrease organic matter, and alter pH.[5]
- As much as possible, ensure that your crops receive adequate water, which helps facilitate nitrogen fixation and mineralization. Ensure that your irrigation system is working and the soil is mulched where possible.

Using Cover Crops for Fertility

Cover cropping is a method in which growing spaces are taken out of production for a time to allow grasses and other plants to replenish the soil's nutrients. It is to a small degree an imitation of the fallowing techniques used by indigenous cultures worldwide. Though fallowing periods often last years, cover cropping can offer some of the ecological benefits that fallowing periods do in our modern production context. Cover crops can sequester, hold, or gather nitrogen and other nutrients. They can feed microbial life. They can fend off weeds by producing allelopathic compounds. Indeed, when done well, cover crops can be the mainstay of your fertility program—where space allows, of course.

The amount of growing space required to utilize cover crops as a primary component in any fertility program is an important consideration. For example, it would be challenging to establish an effective cover cropping regimen on a quarter-acre plot while simultaneously running a profitable market garden, because many of the most valuable cover crops require a growing period that lasts several months. They need time to grow and often time to be terminated (this process is described later in this chapter). Therefore a cover cropping regimen that would supplant outside inputs requires a substantial portion of the growing space to be out of production for months at a time. On our farm, we manage not quite three quarters of an acre in production, and so we use a hybrid system. We manage the nutrition of roughly half of the garden by growing cover crops, and we manage the other half through use of fertilizing inputs as described previously. The cover cropped beds are where we plant long-season crops such as tomatoes, potatoes, and garlic. Those areas that aren't planted with a cover crop are intensively cropped with shorter-season crops such as lettuce, beets, carrots, and arugula. Every few years we swap the management program for these plots. Also, whenever possible, I try to sneak in cover crops between cycles of cash crops.

Figure 7.10. The nitrogen-fixing nodule from an Austrian winter pea plant. A red color inside the nodule indicates nitrogen fixation.

Different cover crops provide a variety of benefits to the soil, and many annual crops (in fact, almost any crop) can be used as cover crops. Certain plant varieties are especially well-suited to the task. Buckwheat, for instance, is a very fast grower, an excellent microbial gatherer, pollinator supporter, and weed suppressant. Legumes—when the proper rhizobium bacteria are present—are adroit at nitrogen fixation. Cereal rye makes a superb mulch and also produces allelopathic enzymes that suppress weeds.

Some attributes of a good cover crop candidate is one that:

- Germinates and grows fast to outcompete weeds
- Provides biomass for mulch and/or weed suppression
- Can be terminated with tools you have and within the timeframe needed
- Is available as seed that is untreated with fungicides or herbicides
- Grows at roughly the same rate as any other cover crop you pair it with (buckwheat, for instance, is a fast grower and so you would pair it with another fast grower)

There are many great resources available on how to grow and manage cover crops, but few of them are built around the small-scale use of cover crops (as is common in market gardens) and the most effective methods for terminating cover crop mixtures. Refer to appendix A, "Cover Crop Use and Termination Guide" on page 241 for descriptions of the most popular and accessible cover crops, their uses, how to sow them, and how to pair them with other crops.

PERENNIAL COVER CROPPING

A perennial cover crop is a short-growing, nonaggressive permanent plant that covers a garden bed. The rationale for planting perennial cover crops in market garden beds is that these plants photosynthesize constantly, meaning they continually fill the root zone with nutrients and microbial diversity. In theory, a grower can then transplant a desired cash crop into the bed and that crop will have access to all the minerals and nutrients the perennial cover crop has mined. In essence, it provides all the benefits of a cover crop without the hassle of having to terminate it.

Though this is an interesting and exciting technique, it is still in its infancy. This approach needs a great deal more work and research before I would recommend it as a commercial growing technique. In our trials, perennial cover cropping has resulted in poor performance from the desired crop, such as the winter squash growing in a bed of perennial clover shown in figure 7.11. It resulted in very small fruits and nonvigorous plants. We, as well as other growers, are continuing to trial different perennial covers, however, and I hope the future results will be more promising.

COVER CROP TERMINATION

Clearing a bed of arugula versus a bed of six-foot-tall (180 cm) winter rye with little or no soil disturbance are two markedly different tasks. A bed of

Figure 7.11. Winter squash growing in perennial clover.

arugula is easy enough to clear. But a bed of cover crops (often grains and sometimes grasses), because of their density and root structures, require a robust approach to termination. Using a stirrup hoe, for instance, would feel like a fairly daunting option when you're confronted with a dense stand of sorghum sudangrass in August.

Cover crop termination is the area in the no-till system that could use the most development. Simply put, when done right cover crops can be just about the only amendment you ever need—your fertilizer, your mulch, your microbial gatherers, your beneficial habitat. Indeed, cover crops are the best soil amendment we've used. No other soil amendment we've ever added has provided the direct and positive plant response that we see in our vegetables when they are simply preceded by a cover crop. If a cover crop is not properly terminated, however, it will become a weed and compete with your crops. If growers had a larger array of termination techniques for cover crops, it's safe to assume there would be a significant increase in the use of cover crops and a reduction in the use of mined minerals, synthetic amendments, commercial bulk mulches, and other environmentally perilous soil additions. For large farms, the use of cover crops is also arguably more scalable than deep compost mulching in terms of both price and logistics (that is, moving mulch is much more challenging and labor intensive than simply sowing a cover crop).

Each cover crop has somewhat different requirements in terms of termination, and I cover this in detail in appendix A, "Cover Crop Use and Termination Guide," on page 241.

Until recently, the conventional practice for terminating an established cover crop has been one of two things: spray with herbicide or crimp with a roller. Assuming you, like me, do not consider the first one a viable option, the latter has been the only alternative up to this point. That said, this preferred option is not without its limitations.

Roller-Crimping Method

The roller-crimper method of terminating a cover crop, in its essence, is when you utilize a heavy rolling implement to crush and kill a dense stand of cover crop. The challenge with roller crimping on its own is that timing is king, and the challenge with timing is that Nature is king.

Let's consider one example. During a cold spring, the "milk stage"—the stage at which rye and other cereal grains are most easily terminated—may not manifest until weeks later than during a warm spring. To elaborate for a moment, the milk stage is the point right after flowering but before seed set that many cover crops go through. That transition stage is the crop's most vulnerable time, as they are pouring their stored energy into seed

Figure 7.12. A simple mowing is enough to terminate some types of cover crops. Buckwheat in particular—excellent for encouraging microbial populations—is easy to kill without any tillage.

production. This stage only lasts about two weeks or less in most cover crops before viable seed is produced. You can imagine that, depending on innumerable factors including moisture and temperature, timing your crimping with the crop's milk stage can be hard to get right. Crimp too early and the crop won't die. Crimp too late and you replant the crop.

Many larger-scale growers use some amount of herbicide at crimping to terminate the crop. When terminating with a crimper organically, a well-designed crimper, such as the one developed by the Rodale Institute, comes in handy. Used at the right time, a roller with a "chevron" design on it helps tremendously with generating the most effective rolling possible.

All that said, I believe for cover crops to be a truly successful approach to fertility, the whole termination element must be more flexible. There have to be ways to terminate these crops without turning them under when the grower needs them to be gone, not just when the crop is in a vulnerable stage. Many of the benefits of cover cropping are established in the soil before milk stage, especially in overwintered cover crops. In the next pages, I discuss some of the innovations and termination options that have made cover cropping more reasonable on a small scale

Organic Large-Scale No-Till Cover Cropping

Using cover crops as a mulch and fertility system makes quite a bit of sense on larger scales where, for instance, covering the soil with a deep layer of compost for planting may not. I feel that there are many great tips that any grower of any size can glean from those who are producing on larger scales, especially in terms of efficiency and equipment management.

One example is how Shawn Jadrnicek of Wild Hope Farm in South Carolina regularly applies mulch to his cover crops after they have been crimped. The leaves help to hold the cover crop down, and they also aid in bolstering the mulching potential, because they provide more organic matter for the soil. We've been experimenting more and more with adding hay overtop our cover crops after crimping (because leaves are not always easy to get here) to emulate this idea.

Gabe Brown of Brown's Ranch in North Dakota uses a diverse array of cover crop species on thousands of acres. The rainfall there is extremely low, so the addition of multispecies cover crop mixes helps ensure that something will grow well while also improving the diversity of soil life. Brown does not crimp but rather intensively grazes his cover crops with livestock. He then sows cash crops using a seed drill into the stubble and smashed forage. Since learning of Gabe's work, we have started employing many more species to our mixes and occasionally grazing them with sheep. I look forward to the day when someone invents a seed drill that makes sense for smaller-scale farming, though the double disc opener for the Jang seeder provides a similar effect. Look to larger-scale growers such as these to discover what's possible with low inputs and the right equipment.

Crimp and Tarp

Few tools have changed the farming landscape quite like the silage tarp. Indeed, the popularization of silage tarps has absolutely opened doors that weren't open before in terms of low-impact soil management options.

I have already articulated the pros and cons of tarps as well as how occultation can be effective in crop termination. For cover crops the story is much the same as killing any crop with a tarp, though there are some nuances to keep in mind.

I first learned this method of cover crop termination from Daniel Mays, author of *The No-Till Organic Vegetable Farm*, who farms at Frith Farm in Maine. I have had some good success implementing it in various ways on Rough Draft Farmstead. Essentially, cover crops are planted on the bed tops. When it's time to terminate the crop, the goal is simply to get the crop down onto the ground before covering with a silage tarp. The closer the crop is to milk stage, the greater the biomass will be and the easier the crop

Low-Budget Crimping Tools

A nice roller crimper with a chevron design and three-point hitch is a great tool, and it has primarily one use: rolling and crimping cover crops. However, single-function tools are generally not great investments on small farms. So what are some low-budget alternatives?

Any heavy implements, such as a flail mower (not in gear for mowing) or a power harrow, can be pulled over a crop to help smash down the plant tops. Before tarping or solarizing, we often use our harrow—the tines raised up a couple of inches above the soil with the PTO

Figure 7.13. A T-post or board can be strapped to the feet and used to press down a cover crop before covering.

will be to kill. A silage tarp is laid overtop the crimped cover crop and kept there for several weeks.

What I like about this strategy is that the maturity of the cover crop is less relevant than if you were to manage it with a crimping alone. The tarping also adds greater insurance that the cover crops will be terminated than crimping would. Note that a crimping or mowing is requisite (power take-off) engaged—to sort of crimp and rip the cover crop simultaneously without disturbing the soil. It spreads the plant residues nicely. You could also use a bed roller with significant weight added, or roll a barrel filled with water down the bed. Use your imagination here—no doubt you have something heavy around your farm that needs more work to do. Plus with all the other options presented here for completing the crop termination, whatever you use doesn't have to entirely kill the cover crop. It only has to lay it down and render it vulnerable. Of course, if your tool can both crimp and kill it, all the better.

Figure 7.14. I ran this power harrow with the tines spinning above the soil surface to help smash and scatter the plant tops of this winter cover mix before tarping.

for this to work. If you try to drape a tarp over standing cover or plant material the crop will take much, much longer to succumb. When plants are pressed to the ground and covered, not only are they cut off from fresh air and photosynthesis, but digestive soil organisms can begin to break it down. It is therefore worth the extra work of crimping or mowing before tarping.

Crimp and Solarize

Solarization can be a relatively quick termination option for cover crops as long as the timing works out. If a crop has to be terminated in the early spring and you run into several cloudy days, the solarization method is not going to produce the desired outcome.

The keys to success are a good initial crimping and a tight, clear, hole-free plastic sheet set tightly overtop the plant matter. The plastic must be more firmly secured in this situation than your typical bed flip, as the plant matter will hoist it up slightly, allowing for more cool air and wind to move underneath the plastic. Secure the sides and the middle as well as you can. Go overboard.

A chopped-up lettuce crop may take 36 to 48 hours to solarize, but a thick stand of rye and vetch could take five to seven days or even longer, depending on the amount of sunlight. The thick mulch produced by the crops like rye is very insulating. It will protect the soil well underneath, thus slowing down the crop termination. That said, even if it takes a fourth or fifth day, solarization still provides a faster termination than occultation with an

Figure 7.15. This cover crop stand that was crimped then solarized for four sunny days is now ready to plant.

opaque tarp. Once the plastic is removed, wait a couple of days before planting to make sure the cover crop is fully terminated. If not, continue to solarize—to plant into nonterminated cover crops is to plant into weeds.

Avoid buying plastic specifically for the purpose of solarization. As it so happens, the plastic used for a high tunnel must be changed every few years. If you or someone you know has a high tunnel, you should be able to acquire some plastic for solarization affordably. Greenhouse and high tunnel plastic is UV treated and so will last for a long time, even if it is being used in direct sunlight. I have heard of people buying other forms of clear plastic to use for solarization, only to find it disintegrates after a couple of uses.

Figure 7.16. These brassica crops are suffering from competition with sorghum sudangrass that was not fully terminated after an inadequate solarization.

Crimp or Mow Then Mulch

Once the cover crop has been pressed or mowed, another option is to cover it with material such as cardboard or paper topped with compost. This method allows you to replant almost immediately, as described in the "Cardboard and Paper" section on page 117.

In this case, cutting through the cardboard isn't a good idea, because the green cover crop underneath is not an ideal environment for crop roots. Instead, cover the cardboard with a layer of mature compost six to eight inches deep and plant into the compost itself. Shallow-rooted crops such as lettuce, arugula, spinach, or even Parisian Market carrots will grow and produce well in the compost as the cardboard and underlying cover crop break down beneath. Alternatively, allow the cover crop to dry for a few days before covering it. That will make the situation below the compost more hospitable to plant roots, especially if the next crop (such as tomatoes) must be planted fairly deep.

Weed Whacker

A weed whacker with a bush blade can be employed to terminate cover crops at most any stage, though if not cut low enough, or not at the milk stage, the covers will regrow, especially grasses. The blade should be sharp. Point the edge as close the base of the plant as possible and slowly pass back and forth across. Though this method is very physically demanding, the weed whacker is a nice option for retaining the thick straw mulch of the cover crop and clearing a bed at the same time. However, I would not recommend trying this on beds that are more than 100 feet (30 m) or 200 feet (60 m). Other methods may be more efficient in larger scale beds.

Sickle Bar Mower

Most often equated with mowing hay, the sickle bar mower could see a resurgence in coming years on the no-till market garden scene.

It is more difficult to terminate certain cover crops with this tool when said crop isn't at the milk stage, but when set low enough, a sickle bar mower has a fair amount of potential for cutting and mulching beds simultaneously. The resulting mulch is left on the surface, baled, or pulled into the pathways. The choice depends on the intended succeeding crop. This implement is generally height-adjustable, so you can cut very low to the soil surface (less robust sickle bar mowers are prone to clogging, so keep that in mind). After mowing, always wait to make sure the crop is fully terminated before planting. One week should be sufficient.

Winter-Kill

If the practical side of creating living soil has a motto, it is perhaps, "Don't do any work that Nature will do for you."

As it so happens, Nature has several great ways to employ crop termination.

The classic winter-kill cover crops are peas and oats. These are planted in the late summer, allowed to mature (preferably not all the way to flowering), and then they are killed during the winter months after a few repeated freezes. The advantage here is that they provide a light mulch for the winter and feed the soil in preparation for spring crops. In our climate here in central Kentucky, we struggle to get the oats to winter-kill some years. We have found that peas mixed with cilantro, turnips, and radishes are a better mix for winter-killing than oats.

Alternatively, summer cover crops can be used, such as sunflower, sorghum sudangrass, cowpea, and sunn hemp. These are planted later in the summer (for us, around mid-August). They grow quickly through the remainder of our summer. These sensitive crops are easily killed in the first few frosts, leaving a substantial mulch behind.

The beds are then ready in the spring to be planted or to have the mulch raked to the side in preparation for direct seeding.

Steam, Heat, and Flame

Using some form of heat or steam to terminate cover crops after crimping is an area that could use further exploration. In our trials terminating buckwheat by crimping and then flame-weeding we found that the top layer of green mulch protects the bottom layer. As a result, the bottom layer is not killed and eventually regrowth pokes back up through the killed surface material. That doesn't mean this method is unworkable, it just needs some tweaking.

In order for this method to be successful, I suspect that the mulch would have to be moved around with a rake or tines—preferably while the heat is being administered—so that all the stems are briefly exposed to the flame. I imagine the same would be true for steam or for some sort of heated roller. More trials and tools need to be created for this to become a viable option.

Bed Preparation without Tillage

In a bed of annual vegetables, the majority of plant roots do not extend as widely and deeply as roots in a bed of perennials. Therefore, the bulk of the microbial life in that bed is located in a very small area just below the surface. This area is called the rhizosphere. This rhizosphere is also where most of those microbes hang out, and most of them won't travel far for minerals to barter with living plant roots. So, ideally, any fertilizing amendments, including microbial amendments, should be added somewhere near the root zone to allow microbes to easily access them. Getting them in there without tillage can be accomplished in several ways.

INCORPORATING AMENDMENTS WITHOUT TILLAGE

On our farm we begin by simply applying any amendments evenly to the soil surface and then lightly raking them in before seeding or transplanting. The idea here is that as we run a seeder through the soil or transplant crops, the implement or transplant root balls push any applied amendments into the zone where roots will grow. Results with this method have been largely successful as long as our amendment mix is indeed spread as evenly as possible—the measurement of success being vigorous and relatively even crop growth. I have begun to prefer a tine rake to a bed rake, because of the tine rake's thin, flexible tines. The tine rake is gentle and moves less surface mulch or soil than an average bed rake. The tine rake, however, does not work as well as the bed rake for forming a seedbed. We either form the bed first and then rake in amendments or we spread and rake amendments first, and then smooth out the surface of the seedbed by flipping a bed rake upside down and dragging the flat side of the rake across the bed.

Another method we utilize is to simply bury the amendments with compost mulch before planting. We spread amendments onto the bed surface and then add a couple of inches of compost overtop. This also has the effect of positioning the amendments in the zone where the roots will be. A variation on this approach is to apply the amendments, broadfork the bed, rake, and then mulch with compost before sowing or planting.

Many growers use a tool called a tilther—or a similar implement that attaches to a rototiller called a precision depth roller—to incorporate amendments. These resemble miniature tillers that work the top one to three inches of the soil. It is similar perhaps to a rake, but mechanized (see figure 7.17). Their resemblances to the common tiller make them controversial among the no-till community, but we have trialed the tilther at our farm and found the depth easy to adjust for a very shallow "stirring" that is not dissimilar to the action of a handheld rake. A tool like this can incorporate a mineral and nutrient mix more evenly than some of the other methods described in this book. On our farm we avoid tools such as the tilther only because they tend to expose more soil and bring more weed seed to the surface. But they can be a fast and effective means of incorporating amendments with minimal disturbance while creating a leveled seedbed all at once.

Another approach to applying amendments without tillage is to spray nutrients or minerals that have been diluted in water onto compost before

Figure 7.17. The tilther is a very low-impact tool used by many growers and its action can help incorporate a small amount of native soil for better seed germination.

adding it to gardens, or to spray the solution directly on the garden soil itself. This process can take many forms, but essentially it involves diluting the recommended amendments in water and then applying them with a large hose. At Assawaga Farm in Connecticut, for instance, they fill a five-gallon bucket with the diluted amendment. Then, using a brass Hozon siphon and a large hose reel, Assawaga farmers Yoko or Alex walk up and down the bed spraying the nutrients as needed to specific areas. In *No-Till Intensive Vegetable Culture*, grower Bryan O'Hara describes a similar method of spraying solubilized amendments onto soil and compost. The idea is that water's ability to penetrate soil makes the amendments more accessible to the microbes, spreads the nutrients more evenly, and gets them deeper into the soil.

For side dressing crops—that is, fertilizing long-season crops such as tomatoes that utilize a large amount of nitrogen—I use the EarthWay seeder with the "pea plate" attachment to push between crops and slightly bury amendments at the root zone (figure 7.18). EarthWay makes an attachment specifically for applying amendments, but I have not tried it. The pea plate works well for my purposes. Side-dressing can give a new life to crops that are long season.

Ultimately, however you choose to amend the soil, the goal for adding amendments is to place them just enough below the surface. This location

Figure 7.18. Using the EarthWay seeder to side dress some cabbages before they create heads with a mix of alfalfa and feather meal.

means they won't be lost to the wind or excess rain but will be close enough to the root zone so that the microbes and plant roots can access them.

PREPPING A SEEDBED

Direct seeding is a needy process. A transplant can often be placed into a bed that is not perfectly groomed, but a seed planted in the same bed is less likely to germinate or thrive. Moreover, because seeds can take several days to germinate, if you run into a crop failure, or a spotty germination, you can be set back several days or weeks. Failures like that can be expensive. Proper preparation of a seedbed, therefore, is paramount.

Most types of seeds that a grower direct seeds into the soil need four elements to germinate: oxygen, moisture, temperature, and (for some crops) sunlight. Plus, all of those elements must come in the proper proportions.

For example, seeds need water to germinate but they will not germinate, or at least not thrive, in conditions that are too soggy and anaerobic. Seeds also require oxygen—which they use to unlock the food stored inside the seed embryo—but will not be able to grow if conditions are too oxygen rich (as they are in chunky compost). If there is too much air around the seed, the meristem—the actively dividing cells that form new root and plant tissue—cannot root into the soil or the seed will dry out and the seedling will fail.

Temperature, for its part, is the signal that tells the seed when to germinate or not germinate. For instance, if you're a tomato seed you do not want to be germinating during the winter—you would not be likely to survive. When the temperature around you is too hot, too cold, or fluctuates too much, then you, the seed, are going to wait until conditions are more optimal. Seeds that require light to germinate (parsley, for example) will never sprout if they are buried too deeply in soil.

Understanding and paying attention to these elements is central to ensuring seed germination and plant success. Here's what that looks like in a no-till context.

No matter whether you are just starting out a new bed or flipping a bed that's already in production, the planting area you wish to sow should be free of weed seeds. Grasses and the like can swallow a bed of carrot seedlings or make a bed of baby greens unharvestable. Cultivate out any potential or sprouted weeds as the first stage of seedbed prep.

If there is any degree of compaction—if you cannot easily slip your fingers into the soil, for instance—consider lightly broadforking the bed to ensure there is proper oxygen for seed germination and growth. Conversely, if there is an existing layer of, say, compost mulch, then the reverse could be true. You may need to lightly compress the soil with something heavy, such as a bed roller, so the seed and its microbial friends can more easily occupy

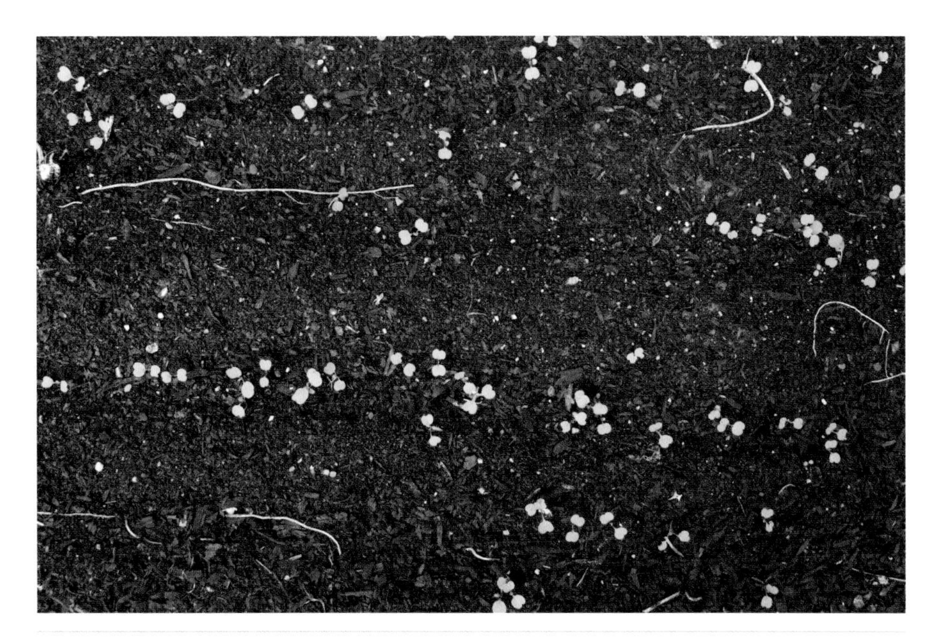

Figure 7.19. Spotty germination occurs when mulch is too deep—as it was with this arugula—when moisture or temperature are not maintained, or when there is too little oxygen in the soil or growing medium.

the space. (Remember, if bacteria and fungi cannot adequately build aggregates between the soil particles because the particles are too far apart, the plants and seeds won't succeed.) Compressing the soil particles together with a roller will allow for more microbial aggregation and movement. If the soil still feels extremely loose to the touch and easily cascades out of your hands when you pick it up, then some amount of balancing of the physical nature of the soil (often called physics balancing) is in order. Adding peat moss, well-decomposed compost, or a similarly textured material at this point will help create the physical texture that the seeds require.

As described previously, some market gardeners use a power harrow, a tilther, or a precision depth roller to smooth out the surface of a bed, making it both level and ready for seeding. A rake can do this job as well to an extent, though a little more slowly and unevenly.

If there is an existing layer of thick, dry compost mulch on a bed we want to seed, we lightly rake some of the mulch off the top. Indeed, if it is too chunky, raking off a small layer does not greatly harm the fungal populations—they were likely not able to form well in that environment anyway. Rather, raking some off allows for better germination of the seeds, and growing plants are always the best thing for the soil.

Soil temperature can be slightly increased or decreased through the use of tarps—white side up can help to keep soil cooler, which is helpful for

Figure 7.20. Using the white side of our silage tarps we can germinate crops more consistently, including cover crops as well as more heat- or humidity-sensitive crops such as carrots or beets.

germinating carrots in the summer; black side up can help to warm up soil in the spring, encouraging both germination and microbial activity. If the tarps are not removed before germination, they will effectively terminate the crop—err on the side of caution. If you put an opaque cover over seeds to encourage germination, always wait until the afternoon after the sun is going down, or for a cloudy day, to remove the covering. Otherwise, sudden exposure to strong sunlight can scorch the young seedlings. Row covers, and to an extent paper mulches, can be used for these same purposes.

Once you have beds prepared with amendments and have created the right conditions for germination, it's time to plant. In the next chapter, I take an in-depth look at tools and tactics that produce seed sowing, transplanting, and crop development results that are reliable enough to stake your living on.

CHAPTER EIGHT

Transplanting and Interplanting

Some amount of soil disturbance is a nearly inevitable part of agriculture, but that doesn't mean our disturbances have to be detrimental. Certainly, we can disturb the soil in ways that hobble it, releasing carbon stores and creating compaction or causing erosion. But we can also disturb the soil in beneficial ways that increase its biological diversity, water-holding capacity, structure, and organic matter. And in my experience, the occasion that offers the greatest potential for beneficial disturbance is planting time, when we are filling the soil with photosynthesizing plants as much and as fast as possible.

This chapter explores the basics of growing great transplants. I will also examine effective seeding techniques and consider nuanced planting practices, such as intercropping and relay cropping, in which multiple crops occupy the same bed at the same time. The goal is to equip you with new tools for keeping the soil planted as much as possible with crops that generate farm income and feed the soil.

Growing Healthy Transplants

The healthier the crop going in, the healthier the crop coming out. Keep that maxim in mind when transplanting crops—even the most robust soils in the world can do only so much when furnished with sickly transplants.

Growing healthy transplants starts with high-quality seeds. Purchased seeds should come from a trusted purveyor. With seed from an unknown source, the grower runs a small risk. Such seeds may carry embedded diseases, such as black rot in brassicas or bacterial spot in nightshades. Poor germination or excessive off-types are other risks of procuring seed from less trusted sources. If you're new to the area where you are farming, or are

Figure 8.1. At this stage of maturity, these healthy lettuce transplants will be ready for harvest only a few weeks after field planting in the spring or summer.

new to growing in general, ask other local growers whether there are any local seed purveyors they use. I absolutely endorse supporting small local seed purveyors, but perhaps start with small orders and run experimental trials to see which purveyors offer the greatest consistency and quality. Avoid falling into the trap of "only heirlooms"—hybrids are simply the vigorous mutts of the seed world, and they can perform excellently and taste incredible.

Germination success also hinges on where seeds are stored between purchase and planting time. Indeed, growers can easily ruin good seeds (read: I have ruined many good seeds). Seeds should be stored in a cool, very dry place. Dryness is more important than temperature. Keeping seeds at room temperature is fine as long as they are kept dry. We have heavy humidity here in Kentucky, so we stash our seeds in plastic containers on a shelf in our walk-in cooler. Freezing is not generally recommended, because to do so effectively the seeds must be dried to below 10 percent moisture before freezing. Otherwise, the combination of moisture and very cold temperatures will reduce germination efficacy. Generally speaking, if temperatures and humidity fluctuate too much, germination rates become less reliable. Seeds of certain species, such as allium or other umbels, are naturally short-lived. Germinations rates dive if the seeds have been stored for more than six months, even in good storage conditions. Protect seeds from pests while in storage—nothing germinates worse than a seed with the germ eaten out of it.

SOIL MIX

The quality of the soil mix is of the utmost importance. A poorly constructed soil mix can reduce germination, vigor, and overall transplant success. A well-made soil mix, on the other hand, can promote consistent germination and growth. It can also sustain crops in small containers (cells or soil blocks) for a longer period of time.

There are several niche soil mix producers, including Vermont Compost Company, Tilth Soil, Earth Care Farm, New England Compost, and Dirt Hugger, who make exceptional products for seed starting. I am becoming aware of new producers every year. It is possible to make your own seed-starting mix, but I find that it is worth the money to simply purchase a well-made product. The mixes I have made myself are rarely of the same quality as what's on the market—the net effect being transplants that are less healthy and vigorous.

CONTAINERS

There is some debate as to which is better for plant health: cell trays, soil blocks, or Winstrip trays. It is generally agreed that it's not ideal to have root-bound transplants—plants whose roots wrap around and around inside the planting container, creating a web that completely encircles the growing medium—because they do not establish well. To be sure, this wrapping occurs in cells—and to a significantly lesser extent in Winstrips—but it does not occur in soil blocks. We prefer soil blocks and Winstrips on our farm for ease of transplanting and the vigor of the crops, but we have grown plenty of healthy crops from transplants started in cell trays, as well. What little

Figure 8.2. These pepper plants at Pavel's Garden near Louisville, Kentucky, are growing in soil blocks made with Fort Vee from the Vermont Compost Company. Quality potting mix makes an enormous difference in plant vigor and success.

research data I can find largely backs up my experience, suggesting there is little notable difference between plant performance in blocks and cells. However, more studies should be conducted to understand what, if any, advantages or disadvantages there are among containers in terms of establishment, yield, disease resistance, pest susceptibility, transplant shock, vigor, and other parameters.

ENVIRONMENT

In a greenhouse, as in the field, basic environmental conditions—including moisture, temperature, light, and depth of planting—play an important role in the seed-starting success. Some seeds have particular needs for germination. Lettuce seed, for instance, goes dormant above 80°F (27°C). Celery requires light to germinate, so it cannot be germinated in a dark area. (Refer to your seed purveyors for the best individual advice.)

Many growers use germination chambers to increase consistency of humidity and temperature—these can be constructed out of old freezers using heating plates and thermostats. Once seeds have begun to germinate, they will need immediate sunlight or they will become "leggy"—the seedlings stretch to reach sunlight, creating a long, weak stem. Transplants with leggy stems struggle to hold themselves up in the field, which can lead to disease if the plant lays against the ground, among other issues.

Think of every tray in your greenhouse as if it were a small, mobile garden and treat it accordingly. As in garden soil, consistent moisture in the soil block is important throughout the transplant's life. Every day, touch the soil around the plants to confirm that the plants are not drying out. Even when the blocks look wet on the surface, the soil at the root zone can be dry, stressing the plant. Overwatering can be just as devastating as underwatering. The soil should feel wet but not soggy. I encourage you to explore bottom watering options for your greenhouse, as this will reduce the moisture on the leaves, which can invite disease. Bottom watering is simply having a system of soaking your growing trays from the bottom—you just need something like a tray or table that can be filled with water. There are also fabrics called capillary mats that absorb water out of a reservoir. The planting trays are then set on top of the mat or are placed in the trays of water. They then absorb the water and keep the young plants hydrated. We have found this strategy to be a much easier way to manage moisture for many crops, especially large tomato transplants, which tend to be extra thirsty. Just make sure the plants are not sitting in standing water for more than an hour or so at a time.

Hardening off is the process of gradually subjecting young transplants to periods of direct sunlight and fluctuating temperatures over the course of several days. This process helps prepare them for the harsher environment of the garden where sun and wind are not mitigated by greenhouse plastic. Any plant grown under a grow light will require at least a week of exposure to direct sun to be hardened off before transplanting.

TRANSPLANTING

To ensure the presence of vital biology right from the beginning, we always employ compost tea or extract at or before transplanting. On planting day, we fill a tray or tote with compost extract at about one inch (three cm) depth and allow the plant roots to soak it up for a few minutes. If they are large transplants like tomatoes, we saturate the plants in trays filled with compost extract. We also fill each planting hole in the growing bed with compost tea or extract before setting the tomato plant in place. Since we have to make a hole in the soil in order to plant—that is to say, we have to "disturb the soil"—this is an excellent opportunity to turn that potential negative into a positive by introducing the best biology possible right at the root zone.

Avoid transplanting when the soil temperature is dramatically different from the temperature of the soil block. For instance, plunging the warm roots of a tomato transplant into cool soil can stunt the tomato plant's establishment, making it susceptible to pest and disease. Spreading a tarp or dark mulch over a growing bed in the early spring before planting time can help warm the soil. Light-colored mulches and cover crop residue help to cool down soil that is warmer than desired.

I've said it before, but it's worth repeating here: Moisture is essential for biology to thrive and to help your plants thrive, too. Plants should be immediately watered-in to ensure fast establishment. A deep saturation following transplanting makes a world of difference in crop success; consistent moisture throughout the plant's life is also important.

Basic Interplanting Strategies

Nature is not a huge fan of monocultures. Or if she is, she's awfully quiet about it. Rarely in Nature do you find a single crop growing on its own (beyond the giant aspen groves of North America, which are not monocultures but gigantic individual organisms). Therefore, if mimicking Nature to the best of our abilities is a primary goal of no-till market gardening, then we should look at ways to diversify the crops that occupy our beds.

As I write this chapter, I want to emphasize that there are some beds in our gardens planted with single crops, but there are also dozens of beds shared by two, or even three, different species. Perennials and annuals are mixed together. Flowers and vegetables are together, onions come up under squash, beets grow among ginger. It's become a passion of mine to find the

FOCUS

No-Mulch Methods

The focus of a no-mulch method or no- or low-tillage market gardening is on managing the first inch or few inches of the soil in the permanent beds and pathways. This is arguably the most common form of soil management for no- or low-till market gardening that I see. Though crops may be mulched from time to time in this system, mulches are not the focus. Weeds are generally managed through well-timed plantings and light surface cultivation, often with very specific and precise tools. Water is managed through irrigation. Sometimes the physical nature of the soil is balanced by adding amendments like peat moss and through mineral balancing. Compaction is generally mitigated using a broadfork.

The no-mulch method is a more technical form of soil management than using a mulch covering. Without mulches the weed pressure in beds requires consistent attention, thus necessitating more cultivation time. However, crop germination rates and plant performance are very reliable, and can be highly productive in

Figure 8.3. Healthy crops can be grown without mulch, but the trade-offs are a greater need for weed cultivation and lower soil moisture retention than in mulched systems.

no-mulch scenarios. Growing without mulches is also the most common form of growing system in which tools like the paper pot transplanter are successful.

On the negative side, due to the shallow stirring of soil, garden beds in a no-mulch system are more likely to have weed pressure. This means they need an established cultivation regimen. Beds without a mulch covering are also more susceptible to erosion and compaction in substantial rain events. Lower moisture retention may also be an issue in more arid climates where water is inaccessible or expensive. Specialized equipment and hand tools are required to make this efficient, thus increasing the up-front capital required.

The groundwork for this sort of system was laid by Eliot Coleman, and it is described in his classic book *The New Organic Grower*. His daughter Clara Coleman is continuing to build on this system at Four Seasons Farm in Maine. Jean-Martin Fortier and Maude-Hélène Desroches further refined this model on their efficient and profitable one-and-a-half-acre farm in Quebec, Les Jardins de la Grelinette. Fortier authored the pivotal *The Market Gardener* and has further proven the efficacy of this system on seven acres at the nearby La Ferme des Quatre-Temps outside of Montreal. Another prominent proponent of this style of marketing gardening is Conor Crickmore. He and his wife, Kate Crickmore, have developed a highly profitable version of this system, relying heavily on a tilther for soil work, at Neversink Farm in the Catskill Mountains in New York State. Crickmore introduced me to the idea of physics balancing—balancing the physical nature of the soil through the addition of carefully calculated quantities of amendments and materials like peat moss.

right combinations and the right timing for growing multiple crops in the same growing space at the same time—that is, to interplant. But, of course, sometimes it doesn't make sense, and that's okay, too.

Interplanting is nothing new in agriculture. Throughout agricultural history, interplanting, (also known as polyculture or multicropping) has been the norm. Many indigenous cultures still emphasize interplanting in their farming practices. Modern conventional agriculture, however, abandoned that model for the seductive simplicity of monocultures—market gardening not excluded. The reintroduction of interplanting practices on commercial farms in the United States began with renewed interest in organic farming and gardening in the middle of the twentieth century. It has gained popularity with market gardeners over the past decade, and it is even attracting interest of some large-scale commodity crop growers.

What traditional agrarians understood was that monoculture farming is inherently depletive—depletive of nutrients, labor, microbes, and soil. Studies of native landscapes have demonstrated higher productivity from plants grown in polycultures as compared to plants grown in monocultures. Polycultures of green onions with cucumbers increase the nutrient accumulation and biomass of the cucumbers versus when cucumbers are grown in a monoculture. An accounting of research on the benefits of polyculture is beyond the scope of this book, but based on my reading of that research, and on my own farming experiences, it's evident to me that well-designed polyculture systems offer more to the soil and grower alike by decreasing pest populations while increasing soil biodiversity, nutrient uptake, and overall yield in the best executions.

If you haven't yet tried interplanting, I advise you to venture in cautiously. In fact, I insist in doing so myself. Experiment often; but if planting a single crop makes sense at a particular time and in your context, it's okay to skip an interplanting. The "as possible" mantra of no-till is as applicable here as anywhere. Here, I offer some guidance and some proven interplanting schemes, but I encourage you to run small trials of your own. Take good notes. Interplanting can be very effective but it can also have negative effects on yield and plant health if misengineered. Be observant and thoughtful when venturing into this realm of market gardening.

FILLING UNUSED SPACE

Interplanting excels as a way to maximize growing space in the market garden. An example of this is growing a round of radishes or lettuce below young tomatoes. Tomatoes require a couple of months to fully occupy a bed. While they grow, an interplanted crop not only provides diversity and as much photosynthesis as possible to a bed, but it contributes a financial boost for an operation. Oxton Organics, who has been growing organically since 1986 and are one of the longest running organic farms in England, grow agretti below their tomatoes. We've found that beets and green onions also flourish in that spot.

Other examples of maximizing bed space include green onions beside peppers, beets beside eggplant, lettuce below sweet corn. The latter example demonstrates the idea of boosting the value of slow-growing, less profitable crops like sweet corn, okra, peanuts, or edamame. If you can fit one or two hundred head lettuce plants below those plantings, it perhaps makes those crops more financially reasonable on small acreage. This is discussed in more detail in the following sections.

FILLING IN GAPS

One of the most practical interplanting recommendations I've received comes from Jared Smith of Jared's Real Food. Jared plants lettuce heads into gaps in a carrot bed when the carrot germination is poor, as shown in

Figure 8.4. Planting lettuce to fill the space below tomato plants early in the season maximizes the production potential of the growing area as the vines mature.

Figure 8.5. Lettuces grow happily in the gaps between carrot tops in a stand of fall carrots where germination was inconsistent.

figure 8.5. To explain, carrots can be slow to germinate, so it may be several weeks before you realize that your carrot germination wasn't adequate. At that point, you have to choose between continuing to tend a mediocre stand of carrots or culling it and starting over. Neither is ideal financially. Having an extra tray of lettuce to fill in the gaps makes that decision for you, because the lettuce will help bring in some revenue that is otherwise lost due to poorly germinated carrots.

We have also found this strategy to be so handy that we always keep an extra tray of head lettuce growing just to fill in gaps as they appear in plantings of green onions, beets, or turnips. This strategy requires the grower to relinquish their inclination to maintain a tidy aesthetic. But having a full canopy increases both the plant coverage and weed suppression while maximizing the earning potential of a poorly germinated bed. Note that most head lettuces need at least seven inches (18 cm) of space on either side of their center to head up well without disease issues. Many romaine varieties prefer closer to 10 inches (25 cm) or more. That is to say, if you do want to fill in gaps, make sure each gap is ample enough to accommodate the filler plant at maturity.

STRUCTURAL BENEFITS

Some crops have physical attributes that can offer benefits to other crops. An example of this is tall crops that can provide shade to shorter ones during the summer. We have had success transplanting summer lettuce into a celery bed a few weeks before the celery reached harvest readiness, as in figure 8.6. Lettuce is sensitive to excessive sunlight in the summer, so this little bit of shade allows the lettuce to establish in the heat with some shade before the canopy is opened up. Okra, corn, sunflowers, tomatoes, and pole beans—almost any tall crop can provide this shading effect where space allows. We've also observed that dense plantings of head lettuce can provide a blanching effect to green onions transplanted between the heads, which adds more of the white that customers often seek out.

The most well-known example of crops physically benefiting others is in the ancient practice of planting corn, squash, and climbing beans together (sometimes adding or substituting other crops such as sunflowers for corn). This combination, known as the Three Sisters, was originated by Mayans and other indigenous people of Mesoamerica. Today, gardeners from all over the planet utilize the Three Sisters system because it is such a productive growing method. Three Sisters is demonstrative of the structural potential of interplanting. The large corn stalks provide a trellis for the beans, and the squash provides a canopy to shade out weeds and retain moisture for the other two. (The beans don't give much structurally to the combination, but because they sequester their own nitrogen, they don't take much, either.)

Most natural ecosystems offer abundant examples of diversity and structural mutualism that we can easily replicate in the market garden. Corn, sunflowers, okra, or other tall crops act like trees; shorter crops like beets or lettuce act as the understory. Beans or peas are the vines, and the layered canopy of leaves effectively works to absorb all the available sunlight, preventing excessive heat from burning up the carbon and moisture in the soil. Instead that sunlight is used to create an energy source than can be channeled underground to fuel the soil life.

ENCOURAGING BENEFICIALS

The greater the genetic difference between plants, the greater the composition of their root exudates differs.[1] For example, most annual flowers are of different genera than annual vegetable crops and thus contribute their own

Figure 8.6. Lettuce transplanted below celery in the summer enjoys the bit of shade from the celery canopy as it establishes.

Figure 8.7. We plant entire beds of mixed sunflowers and other annuals, as well as interplanting them in beds with vegetable crops. The flowers are good for beneficial insects and birds, and they can yield beautiful bouquets to sell at market, too.

novel exudates to the soil. However, a flower's benefits go beyond what they offer the soil. They also attract predatory bugs, parasitic wasps, flies, pollinators, birds, and other beneficials. Per recommendations from Daniel Mays of Frith Farm, we tried growing sweet alyssum below nightshades (see figure 8.8) and found that they made for an excellent long-season understory. Dr. Christine Jones has recommended flax and nigella for interplanting and insect attraction. No-dig champion Charles Dowding plants marigolds below his cucumbers to help repel aphids. On our farm we have allowed cilantro, lettuce, and basil plants to bolt in and around tomato plantings because their flowers attract parasitoids so effectively. Herbs like dill can provide not only shade but bring in many pollinators as they begin to flower. Perennial herbs such as lavender, rosemary, and thyme are also abuzz with life when flowering and are excellent crops to include around your production beds where conditions allow. Fresh-cut flowers are highly profitable per square foot, so their benefits are not limited to biology if you develop a market for selling them. A thin stand of flowers such as dill, zinnias, cilantro, and/or cosmos transplanted into gaps of lettuce or squash is one of my favorite ways to provide a little extra shade, income, and beneficials all at the same time. Try for one transplant every 10 inches (25 cm), give or take.

Figure 8.8. Sweet alyssum planted beneath tomatoes and around other crops brings in beneficial insects such as the braconid wasp that lays its eggs in hornworms. The wasp larvae feed inside the hornworm, killing it, then emerge and spin these white cocoons in which they pupate, emerging as adult wasps.

Dill, cilantro, buckwheat, sweet alyssum, chamomile, fennel, parsley, lavender, rosemary, thyme, daisy-family plants, and many other flowering plants attract parasitoid wasps and flies that lay their eggs in hornworms and cabbage worms and feed upon aphids. Buckwheat flowers have even been shown to increase the longevity and fecundity of certain caterpillar parasitoids.

Some parasitoids kill herbivorous caterpillars by laying their eggs inside of them, and the offspring then feed on the caterpillars as they grow. But amazingly, the parasitoid offspring feeding on the caterpillar can affect the composition of the caterpillar's saliva, which can in turn alter the plant's genes. For example, the saliva of the parasitized caterpillar can "tell" certain brassica plants to undergo metabolic changes that make the plant a less attractive host for future cabbage moths. Indeed, the saliva of some parasitized caterpillars has the effect of making the cabbage plant tell moths to stay away![2] Of course, not every interaction between these species will yield the same result, but these phenomena are fascinating nonetheless.

MAXIMIZING PROFIT

In order for growers to be successful using these more ecological practices, each pairing must be financially viable. Luckily, interplanting adds an abundance of value to a bed. Take peppers for instance. In the spring we plant peppers in the field around May 10. We start harvesting the first fruits by mid-July. That means there is a period of 60 to 70 days when those beds are producing no income. However, add two hundred heads of lettuce that will sell at three dollars each to the bed and suddenly that bed will be generating $600 before we harvest a single pepper. It is important to run some trials, do the math, and evaluate your results realistically, but I think you will soon discover that interplanting does not have to be solely about soil diversity. It can just as significantly be about making a living, too.

Figure 8.9. No bed in this fall planting is planted to a single crop; instead the beds are split between two crops to increase the plant diversity.

Research has suggested reductions in weeds and more efficient nutrient cycling in polycultures—both of which help a market garden financially in the long run by increasing yields and improving soil performance. But on a more practical level, growing two crops together in one bed helps decrease your susceptibility to financial loss. Take the pepper and lettuce example: If a surprise hail storm destroys the pepper plants but the peppers protect the lettuce beneath, then the bed will not be a total financial loss.

Conversely, interplanting incentivizes growing less-profitable (but no less tasty) crops like okra or sweet corn. Beds of these types of long-season, low-profit crops can be sown with radishes, interplanted with lettuce or beets, or filled in with cut flowers, adding hundreds of dollars to what are generally unprofitable, slow-to-mature crops. I describe more variations on this concept in the section about relay cropping later in this chapter.

The body of evidence supporting intercropping as a means of pest or disease control is increasing, and it's a practice that can lead to higher profits for the no-till farmer. Several meta-analyses have shown that plant diversification in growing areas can decrease pest populations. The one caveat is that the measured yields in most trials were often lower in the intercropping schemes as compared to monocultures of the same crops. That said, there are also plenty of intercropping trials that have shown yield increases.

For the best yields, choose interplanting combinations that make sense for the season and the space. And remember that soil health is always a more significant influence on the outcome than the crop combination. As the title of a famous book tells us, carrots love tomatoes—but that love won't bear fruit in compacted soil. On the other hand, fennel is notoriously antagonistic as a companion plant, but even so we have had success interplanting fennel with lettuce and celery in beds where the soil was loose and lively. A healthy, living soil always sustains crops better than a depleted or compacted soil. The most important companionship is not necessarily between the plants—though they may indeed benefit from one another—but between the plants, the sun, and the soil life.

SOME PROVEN BEGINNER COMBINATIONS

Tomatoes and peppers are good crops for interplanting experiments because they can handle some competition and their growing habit leaves ample open space for several weeks as other crops slowly mature. In the spring, plant a single row of these taller nightshades down the middle of a bed with a row of head lettuce alongside, 12 inches (30 cm) from the row of nightshades, with heads spaced 8 inches (20 cm) apart. Beets can fill that same role, 4 inches (10 cm) apart in the row. A thick band of radishes is also an option. Or, start some green onion seeds in cell trays, five seeds per cell, and transplant the cells four to six inches (10 to 15 cm) apart in rows on either side, as well. Trellised cucumbers can generally take the place of the

Figure 8.10. Three examples of interplanting. At left, peppers are flanked by lettuce and beets. At center, beans with small red romaine. At right, Salanova lettuce with green onions.

nightshades in this scenario, though keep in mind they tend to fill up space much more quickly than nightshades. They are a bit more sensitive to competition, too, in our experience.

Another easy way to ease into interplanting is just to split beds. Say you plan to plant one bed of spinach and one bed of green onions. Instead of planting a monocropped bed of each, combine them, making each bed half spinach and half green onions. The spinach, in this example, will benefit from the onions' ability to make mycorrhizal associations and the two crops will be ready to harvest at relatively the same time. This split bed idea is perfect for our intercropping workhorses: spinach, arugula, beets, green onions, and lettuce.

Start with those and then slowly work your way into other more advanced interplanting, as described here.

Advanced Interplanting Strategies

To help further refine your interplanting techniques, in this section I provide several strategies and concepts that have aided me on my farm in both creating successful interplantings and understanding those pairings that didn't work.

Figure 8.11. An attempt to plant four rows of lettuce and four rows of carrots in a single bed resulted in poor carrot germination, early bolting in the lettuce, and some disease in both. Carrots are not a good competitor when young.

CRITICAL PERIOD OF COMPETITION

"Critical period of weed control" is not common terminology among most small growers, but some agronomists and conventional growers are familiar with it. It is also called "the shade avoidance mechanism." This is the period in which plants are most affected by competition, and conventional growers consider this in making decisions about when to apply herbicides. But it is also a time that can be harnessed for improving interplanting strategies.

When plants have too much competition, they often put more energy into growing tall than producing healthy yields. For two years I mistook the impressive height of my garlic plants that were interplanted with Austrian winter peas for vigor, neglecting to connect it with the poor garlic yield. The results were tall garlic plants, but small bulbs. Similarly, I found that lettuce too densely interplanted with carrots resulted in very tall but low-yielding lettuce, and smaller carrot roots with disproportionally tall greens.

Plants have a number of different mechanisms for sensing other plants. They can sense the touch of competing leaves and roots. They can sense hormones given off by leaves. And they can sense the light refracted off nearby plants. Whatever the mechanism, if a plant senses too much competition early on, the plant will not produce as well. For us, the concept of critical period of competition has become very helpful in designing successful interplanting combinations—and in understanding why certain plantings *have not* worked. At the time of writing this book, there is not yet a critical period of competition study available for all common market garden crops. I was, however, able to track down critical period information for nearly all of them. Indeed, I am amazed at how much information is available—nearly every crop is a major agricultural commodity somewhere in the world and thus deserving of substantial research. I compiled my findings in appendix B, "Critical Period of Competition and Interplant Pairings," on page 247.

Generally speaking, the negative effects of competition are most apparent in direct-seeded crops. By raising transplants you can often bypass the critical period, which perhaps supports the contention made by many growers that transplanting increases crop success. This rings especially true when it comes to interplanting. Another important caveat is that some weeds may have greater or fewer deleterious effects on crops than competing vegetables. The critical periods of certain crops may also vary based on time of year. Moreover, some varieties may be more susceptible to ill effects of competition than others. Do your homework, pay attention, and take good notes during your trials.

BIOCONTROL

Interplanting for control of certain pests is not a new idea, but the research that backs its effectiveness is very new. For instance, researchers at Newcastle University recently demonstrated that the traditional practice of growing marigolds next to tomato plants "significantly slows whitefly development."[3] Certain varieties of marigolds as well as asparagus exude nematatoxic compounds that kill or deter pathogenic soil-dwelling nematodes. The effects may be most beneficial when these plants are paired with or grown just before nightshades or carrots.

Allium-family plants are of particular interest as pest or disease deterrents. One small study demonstrated how lettuce and onions grown together reduce damage on the lettuce from thread caterpillars as compared to monocropped lettuce.[4] Several studies have shown that when cucumbers are grown with green garlic, the soil community is more diversified. Moreover, the cucumbers create more biomass and are better able to accumulate nutrients in the allium intercropping scenario. Interplanting garlic and strawberries can decrease the presence of two-spotted spider mites in the field by 65 percent. The reverse has also been shown—planting onions with certain vegetable crops may reduce instances of allium diseases. In practice, we have had great success sowing onions below two other cucurbits of note—zucchini and squashes.

Figure 8.12. When green onions are interplanted into lettuce, the lettuce helps whiten or "blanch" the onions, and the onions can add pest protection for the lettuces.

It's important to mention that many plants including cereal rye, field peas, cow peas, and several common cover crops contain allelopathic properties that can deter weed growth. For their part, brassicas—especially mustards, broccoli, oilseed radish, or cauliflower—can offer biofumigation properties that reduce disease and pest pressure in the soil. However, these kinds of properties are harder to take advantage of in an active interplanting than in a cover crop, because the allelopathic chemicals can actually slow the growth of the desired plants. That said, sowing an allelopathic or biofumigant cover crop into longer-season cash crops has been shown to have little impact on yield (though most of the studies investigate impact on corn and soybeans, not short-season vegetables). Thus, interplanting such cover crops with vegetables could be an effective approach for utilizing their weed-, disease-, and pest-suppressant properties—if not to benefit the concurrent vegetable crop, then the following crop. For this goal, and to keep the beds photosynthesizing over the winter, we try to sow our winter cover crops among existing cash crops at the end of August and September (provided the cover crop won't interfere with harvest). This style of interplanting is called relay cropping, and I discuss it more in-depth later in this chapter.

ROOT COMPLEMENTARITY

An interesting agronomic concept is that species can occupy their own distinct niches within the soil and share resources in a complementary way. This is often proposed as a partial explanation of the success of the Three Sisters—the roots of each plant occupy different spaces and "coexist without competing for resources," says Dr. Christine Jones. It is unclear whether, in the Three Sisters example, the beans, corn, and squash are sharing gathered minerals or are simply occupying their own niche areas of the soil. Essentially, because all three plants gather nutrients in a different stratum of the soil, no plant is overly competitive with another.

Let's consider another example of root complementarity. Lettuce is a shallow-rooted crop whereas tomatoes have a deep taproot. That is one reason this pairing works well—the lettuce is not competing with the tomatoes for nutrients in the way another deeper-rooted crop would. The lesson here is that instead of forcing two crops that occupy similar niches in the rhizosphere into competition, it's better to plant shallow-rooted crops with medium- or deep-rooted crops and taprooted crops.

I've put together a small composite of the information available on root depths for the more common market crops. This information can help inform interplanting as well as crop rotations—some crops in the same family do not occupy the same root zones (sunflowers and lettuce, for instance). Note that these categories are general—spacing, soil health, soil compaction, bed width, and crop variety will all play a role in how deep or wide a root system will grow.

Shallow-Rooted Crops

These crops have roots that extend one to two feet (30 to 60 cm) deep.

Alliums, Arugula, Basil, Beans, Broccoli, Bok Choy, Brussels Sprouts, Cabbage, Cantaloupe, Cauliflower, Celery, Chive, Cilantro, Collards, Cucumber, Fennel, Garlic, Kale, Kohlrabi, Legumes (short season), Lettuce, Mache, Mint, Mustard Greens, Onions, Oregano, Parsley, Peas, Radishes, Savory, Spinach, Squash (summer), Strawberries, Sunchokes, Tarragon, Turnips, Thyme

Deep-Rooted Crops

These crops have roots that extend two feet (60 cm) deep or deeper.

Artichokes, Asparagus, Beets, Carrots, Chard, Corn (dry and sweet), Cotton, Clovers, Eggplant, Fava Bean, Grains, Leeks, Legumes (long season), Melons, Okra, Parsley, Parsnip, Peppers, Potatoes, Pumpkins, Rhubarb, Rosemary, Rutabaga, Sage, Squash (winter), Sunflowers, Sweet Potatoes, Tomatoes

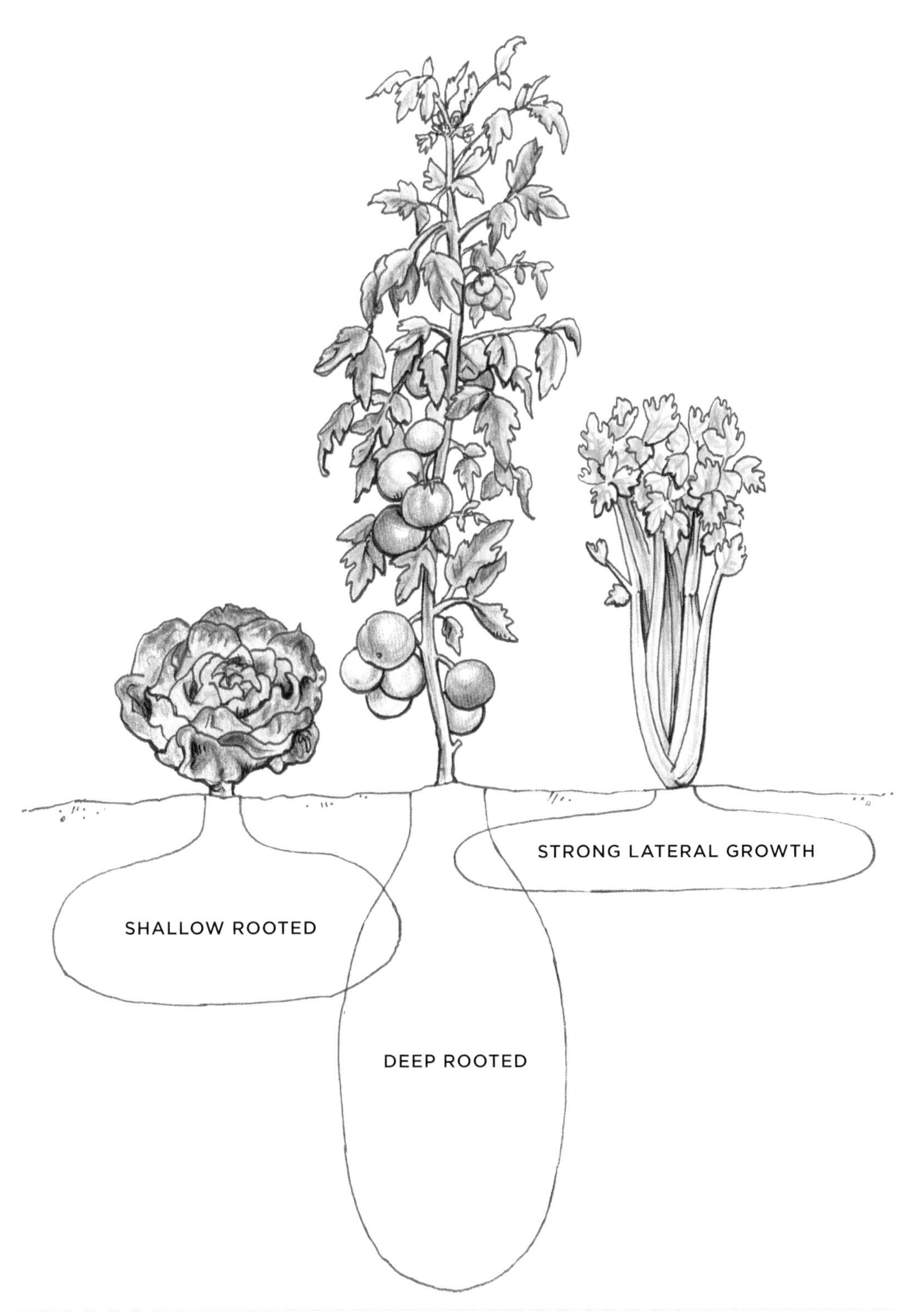

Figure 8.13. Crops with differing rooting habits can be successfully paired together because they will not directly compete for nutrients and water.

Crops with Strong Lateral Growth

These crops have roots that extend laterally three feet (90 cm) or more from the base of the plant. Some are shallow-rooted, others are deep-rooted.

Asparagus, Beets, Cabbage, Cantaloupe, Celery, Chard, Corn, Cucumber, Eggplant, Kohlrabi, Legumes (long season), Okra, Parsnip, Peppers, Potatoes, Pumpkin, Radish, Rhubarb, Squash (winter), Sweet Potatoes, Tomatoes, Watermelon

ADDING PERENNIALS

Perennial plants offer an exceptional amount of potential to the above- and belowground biology of a garden. Whether planted in garden beds, hedges, or at the edges of the fields, perennials help to provide a robust, year-round root system that feeds the soil and gathers microbes and nutrients, often sending roots to depths that annuals can't reach. Shrubs and small trees can provide windbreaks and shade, as well as habitat for beneficial insects and animals. Native perennials are vitally important to native insect populations, as well. As many as 25 percent of native bees in the eastern

Figure 8.14. Mixing perennials such as these blueberry plants into the garden helps retain soil, sustain microbes, add shade, and break up wind, among other benefits.

United States are flower specialists, meaning they rely heavily or exclusively on one species of native flower to survive—many of which are perennials.

Studies have also compared weedy areas at the edge of crop fields with curated perennial hedgerows and found the hedgerows brought in more beneficials than crop pests; while the opposite was true of the weedy areas, which saw more pests than beneficials. The research suggests this occurs because beneficial insects use the pollen and nectar from the flowering perennials in the hedgerows. The hedgerows also provide some beneficial insects with different types of prey. Hedgerows also offer better overwintering habitat than grasses. All that adds up to a greater population of beneficials.

Goldenrod—the state flower here in Kentucky—is a rhizomatous flowering perennial that attracts an incredible number of beneficial insects and predators including the spined assassin bug. And on a more basic level, crucial beneficials such as ladybugs hibernate in tree bark and vines (and houses), so simply having perennial trees around the garden can increase their yearly presence. Then of course many perennial fruits such as blueberries, raspberries, blackberries, oranges, apples, and so on can provide more than just shade and habitat, but a product (or at least garden snacks) to boot. In effect, incorporating perennials—and especially native perennials—into the garden can pay dividends beyond ecology.

For more information about what's appropriate in your region, reach out to native plant nurseries and the Xerces Society.

RELAY CROPPING

One area I find distinctly exciting for maximizing the amount of photosynthetic and monetary potential of a garden bed is a technique called relay cropping. I sometimes refer to this technique as everbed farming, because beds are ever-planted, ever-photosynthesizing. In a normal gardening situation a bed is planted with a single crop. That crop comes out and is replaced by the next. But in a relay cropping scenario, the second crop is planted directly into the first crop so there is no time gap at all between plantings. The bed never stops photosynthesizing.

An example of relay cropping in large-scale grain operations, where it is sometimes referred to as modified relay intercropping (MRI), is drilling soybean seed into standing wheat six weeks prior to wheat harvest. The soybean crop emerges right as the wheat crop is coming out. In regions where wheat overwinters, it can be drilled into standing soybeans around the time they are ready to be harvested. The soybeans are harvested and the wheat holds the soil in place over the winter and provides a marketable crop in the spring. In a market garden context, with more rapidly maturing crops, relay cropping works slightly differently.

In our first relay cropping attempt in the summer of 2018, we transplanted lettuce directly into a bed of celery, as described in the "Structural Benefits" section on page 174. The celery helped provide some shade for the lettuce for two weeks, at which time we harvested the celery. After that, the lettuce began to take over the bed. Before the lettuce fully filled out the canopy, we transplanted green onions from soil blocks in between the lettuces. When the lettuce was mature, we cut out the heads, leaving the green onions to enjoy the bed. Right before harvesting those green onions, we planted our fall kale in the gaps. We harvested kale from that planting all winter. The following spring, we allowed the kale to go to flower to attract pollinators and then to go to seed, which we harvested for future planting. Just before we removed the kale plants with mature seed pods, we interplanted tomatoes.

Another successful example, this one from 2020, involved sowing a bed split between radishes and turnips in March, as shown in figure 8.15. In April we planted zucchini every two feet (60 cm) into that same bed (the gaps were created when we harvested some radishes and turnips for ourselves). We pulled the remaining radishes and turnips in early May, giving the bed entirely to the zucchini. At this point the zucchini were not yet fully occupying the bed so we were able to sow green onions underneath in two thick bands on either side of the young zucchini plants. The zucchini was extremely productive for about nine weeks. We then pulled the zucchini, and the green onions shot up from underneath. Two weeks later we transplanted Napa cabbage at about 12 inches (30 cm) apart in the row (for smaller heads).

Squashes have been the unlikely champion for us in the relay cropping scenario, even with their sensitive roots. We have successfully transplanted summer squash into mature carrots and potatoes and then harvested those root crops without any noticeable ill effects to the squash. In the reverse, in 2019 we sowed carrots below the squash in the same way we did with onions below the zucchini. Once the squash harvest was complete and we cut out the vines, we were able to harvest a healthy carrot crop a few weeks later. The carrots were slightly smaller than expected for the size of the greens—possibly due to competition with the squash during the critical period of competition—but they were healthy and fully marketable.

That said, we've also had some failures (or at least low performers). I transplanted squash and ginger into a bed, and the large squash leaves smashed the young ginger shoots, setting the ginger back a month or so. Sometimes the green onion or carrot crop does not germinate well, gets out-competed, or is too shaded to produce a worthwhile crop (the onions or carrots are exceptional at building mycorrhizal associations, however, so even in failure they contribute). I also attempted to transplant cucumbers

Figure 8.15. A season of relay cropping: In the spring, zucchini transplanted into radishes. In June, green onions germinating under those same zucchini. By September, we have removed the zucchini and transplanted Napa cabbage into those same green onions.

into a bed of maturing onions, but the effort was ultimately unsuccessful (and weedy), though I suspect the issue was compaction in that bed more than the pairing itself.

This relay cropping idea of course can be much simpler than some that I've outlined, such as one technique I picked up from Pam Dawling, the long-time field manager (now retired) at Twin Oaks in Virginia. When Pam appeared on *The No-Till Market Garden Podcast*, she described how she would plant arugula over the winter and then place flags in the bed the following spring to mark where to plant tomatoes. At planting time, she would clear the areas around the flags in preparation for the tomatoes. The tomatoes were then transplanted directly into the arugula, which remained harvestable for a short time longer.

At our farm, we have transplanted tomatoes with overwintered carrots in a tunnel by clearing out spaces in the middle of the carrot bed and then transplanting the tomatoes into those spaces before the rest of the carrots are out. This can also be done with other root crops such as beet, radish, or turnip. The first harvest thins out the canopy, creating spaces into which the next crop can be transplanted. That next crop will go into the newly cleared space and have a few days to establish before the rest of the root crop is removed the following week.

Bed prep for relay cropping is not significantly different than for a standard planting of a single crop. We've learned that compaction must be addressed before the successions start and that moisture must be consistent throughout for both crops to ensure the soil stays loose. We mulch with compost to start, and we fertilize lightly before each crop with alfalfa or feather meal, inoculating compost, humic acid, kelp, and activated biochar. Before each new crop is planted, we again check for compaction—if the soil seems at all compacted, we will lightly pop the soil with a trowel (similar to a broadforking, but in miniature), before transplanting or sowing the next crop. The bed should not have any compaction issues if moisture was well maintained, but it is important to always check. As is usual at our farm, we always leave roots in the soil at harvest where possible, aiming to reduce disturbance for the subsequent crops as they grow in their turn.

Methods such as relay cropping will help you develop a deeper understanding of how different plants grow and interact. Experiment, observe carefully, and take notes. Success with relay cropping will depend heavily on your crop choices, your bed prep methods, your attention to weed control, and your farming goals. That said, relay cropping is an excellent tool for the no-till toolbox, because it adroitly accomplishes all three no-till principles at once: full soil coverage, low disturbance, and a constant supply and diversity of living roots in the soil.

NURSE CROPS

Nurse crops are crops somewhere between relay crops and cover crops. In conventional agriculture, nurse crops are typically grains such as oats and triticale that are used as weed-suppressing crops for establishing perennials such as hay grasses or alfalfa. So, for instance, oats and/or triticale are sowed with alfalfa. Those nurse crops beat out the weed competition for the slower-growing alfalfa, and then the nurse crops are terminated (often with herbicide) or harvested. The faster-growing nurse crops block weeds, and when they are terminated they leave behind room for the intended crop (in this case alfalfa) to take over the bed.

However, in the organic market garden context, the idea of a nurse crop is used a little differently. First, the word *nurse* here can be a little misleading, because it sounds like one crop is directly nourishing another. Though this may happen to some extent—and diversity is generally good for soil health—nurse crops are intended to nourish the soil rather than the other crops, while also suppressing weeds. An example of this is sowing parsley, carrots, green onions, or cilantro into fall brassicas after the brassicas are well established. These nurse crops are fairly hardy winter crops that block cold weather weeds and make mycorrhizal associations (unlike brassicas). Whether those mycorrhizal associations will benefit the brassicas in situ by mining minerals from deep in the soil and bringing those to the rhizosphere is not certain, but that mycorrhizal activity will certainly benefit the next crop in the spring by keeping the mycorrizae fed. What I like about cilantro or parsley in this situation is that you can often harvest it throughout the winter, long after the original brassica is gone. If cilantro, parsley, green onions, or carrots survive the winter in your climate (as they generally do in ours), then they will go to flower early in the spring and provide some excellent food for beneficials.

Legumes such as peas can make a nice nurse crop, and they can provide pea shoots as a salable product, but they don't benefit other crops the way it is sometimes suggested. For instance, a common recommendation is to transplant or sow legumes (such as peas) between crops so that the legumes will share some of the nitrogen that is fixed through their rhizobial association with the other crop. Unfortunately, the research does not clearly support this interplanting benefit and generally suggests the opposite is true: legumes do fix nitrogen, but they keep most of it for themselves. The benefit of that nitrogen fixation can be harnessed via terminated cover crops. The legume is grown and then terminated; this then allows the soil biology to metabolize that nitrogen, eventually making it available for the next crop. If the legumes are allowed to fruit, however, and the fruits and/or plant residue are harvested, then the fixed nitrogen is largely removed from the garden environment. It is possible that because the legumes are using the nitrogen fixed in their root nodules, they may use less of the available nitrogen in the soil, making them a desirable, somewhat self-reliant companion plant.

Nurse crops can be sown with a seeder below the intended crops before their canopy fully fills in. Simply sow as many rows as you can so that the germination is thick. You can also broadcast the seed overtop the main crop and lightly rake the soil to encourage seed-to-soil contact. I recommend being heavy-handed in your broadcasting, as they will not germinate as well as direct-seeded crops. Something that is slow to establish such as parsley can be sown at the same time that the primary crop is transplanted. Again, these are fairly winter-hardy crops so a plan for spring termination must be in place.

CHAPTER NINE

Seven No-Till Crops from Start to Finish

Because the CSA model has been our main marketing strategy for the past 10 years, we have grown hundreds of different crops—all over the tillage and no-tillage spectrum—to satisfy the diversity requisite in keeping the weekly vegetable deliveries interesting for our customers. But here, I have chosen seven crops for in-depth coverage from starts to harvest. Some of them—such as lettuce—are a natural fit for the no-till market garden model, but others—carrots and sweet potatoes—might confound those who aspire to farming without tillage.

To be sure, our methods here in central Kentucky—a rather humid USDA Hardiness Zone 6b with 55 inches (140 cm) of rain on average with some recent years exceeding 70 inches (178 cm)—may not be how you can grow these crops where you are, be it in eastern Canada or western Texas, or even just down the road from me. Your context—ranging from climatic differences to your marketing outlets—may be similar to mine, or extremely different. A bunch of carrots that I sell for three dollars may bring in only two dollars in, say, rural Kansas, but sell easily for five dollars in New York City. And you may not be able to grow sweet potatoes at all because of your climate conditions, or because you don't have enough space to accommodate their vigorous vines and long season. Such differences aside, I know from personal experience how helpful it is to learn about the spacing regimens and growing techniques for specific crops that other growers have devised and succeeded with. Even if I can't copy their practices exactly, their example can lead me to new ideas and adaptations that will bear fruit on my own farm.

In terms of marketing, keep in mind that Hannah and I do not sell any produce wholesale other than some restaurant sales and one aggregated CSA for low-income families. My marketing tips are largely centered on selling at a farmers market and through a CSA. Our farmers market is open for 6 hours on Saturdays and 4 hours on Sundays. Even with the shorter sales day, Sunday is the better market for us because, as at many farmers markets, Saturdays tend to be more of a social gathering. For many people who come to the Saturday market, buying produce is not the main draw. Josh England, our market manager, estimates that at peak season there are three to four thousand customers coming through on Saturdays and two to three thousand on Sundays.

One important note about our marketing system. At the farmers market, we sell all of our produce at one item for three dollars or two items for five dollars. We base our entire production system—from the varieties we choose down to the size that we harvest the vegetables—around that pricing system. I reference it several times in this chapter. This marketing strategy achieves several things:

- Makes shopping easier on the customer
- Makes giving change easier on the farmer
- Speeds up transactions by eliminating the need to weigh or calculate prices

As you read through the crop descriptions in this chapter, notice that I have not focused solely on what works for us, but also on what *hasn't worked*. Failure can be just as illuminating as success—sometimes more so. For each crop I describe, I give the varieties I use, the amount of seed or plants needed, bed prep, planting, harvest, marketing, and even some notable failures. Also note that our fertilization routine for all crops conforms to what I describe in chapter seven, unless specifically denoted.

To follow is a list of our favorite varieties of each crop. That said, we are always trialing new varieties, and I recommend you do the same. Varieties can become unavailable or the quality of a variety can change if a major seed supplier has a bad production year. It helps to have backups. We are also working to breed more varieties of our own that are adapted to our climate. In an interview for my podcast, grower Nancy Kost of The Buffalo Seed Company in Kansas told me about the philosophy she learned growing up in Bolivia. In the region she's from, people don't think of farming and seed saving as two separate jobs—they are one and the same. I believe that as growers we need to work toward that mentality. Most of the seeds we purchase are grown in regions that may not share our climates, or growing methods, or anything else that may affect the performance of a seed, and so the seeds may not be as effective for us as they could be. To improve the efficacy and sustainability of our crops, learning the craft of seed saving is critical. The how-to of seed saving is a topic worthy of a book in itself, and I list some good resources for seed saving in the resources section of this book. And I will continue to interview seed producers for the podcast.

Carrots

Because carrots are such a hard crop to produce consistently in a tillage system due to weed pressure, we have found a niche in our market for selling large piles of carrots every week at market without competition. Carrots are also, notably, a farm and family favorite to grow and eat.

VARIETIES

In the spring we sow Atlas (a Parisian Market carrot type), Romance, and Mokum. In the summer we sow Atlas, Bolero, and Yaya. For the fall we sow Bolero, Mokum, Napoli, Atlas, and Deep Purple. Despite the demand at market for rainbow carrots, we mostly stay away from them because we have found their performance iffy.

ESTIMATED SEED QUANTITIES

We purchase 1,200 carrot seeds per 100-foot (30 m) row. This seeding rate may not be precisely right for your context. This ratio is dependent on the brand of seeder and settings. Note that carrot seed loses viability quickly. For that reason we split our seed orders between spring carrots and fall carrots, placing two separate orders. We do not buy pelleted carrot seed—"naked" carrot seed performs better in our experience.

BED PREP

Attention to bed preparation is essential for carrots because they are finicky about germinating. Also, we aim to harvest a bed of carrots and replant that bed all on the same day. We begin by feeling the soil for compaction. Sandy or well-developed soils are unlikely to be compacted, but our soils are dense and if we detect any compaction at all, we broadfork the bed either before sowing or at harvest time. I prefer to broadfork before sowing. The harvest process can unfold much more quickly when we do not have to fork on harvest day. When we fork before planting, the carrots come out of the ground easily and I can rapidly replant the bed, which is consistent with our goal to minimize the amount of time that a bed is not filled with photosynthesizing plants.

The most important aspect of bed preparation for carrots is that the soil must have good moisture-holding capacity. To germinate, carrot seeds must easily be able to retain moisture for 7 to 14 days. During that time, the seed must be surrounded by and in good contact with soil (or compost) that holds moisture. We check the moisture content of the beds in several places by hand before seeding, just feeling for moisture at the depth of seeding. If the compost feels particularly dry we add some peat moss to the beds using

Figure 9.1. We start harvesting baby carrots when they reach roughly six inches (15 cm) in length, depending on variety. Harvesting smaller carrots also allows us to turn those beds over to another crop much faster, adding to the yearly profitability of those beds.

a Landzie Compost and Peat Moss Spreader. Just one layer before sowing will help retain moisture. Or, if the compost is composed largely of rotten wood chips, as ours usually is, we lightly rake a small amount of the compost mulch off the beds. This leaves behind the more decomposed compost layer below, which better holds moisture. We then seed and water it in extremely well using overhead irrigation.

WEED CONTROL

Many farmers do not bother growing carrots because, if the timing of planting and cultivation is off, fast-growing weeds almost surely consume the slow-growing carrots. Growers often mitigate this weed issue through stale seedbedding—that is, prepping a bed as if you are going to plant it, leaving it for two weeks, lightly cultivating out the weeds, and then sowing the carrots. Flame weeding has also become a popular method of weed control—sowing the carrot seed, waiting a few days, and flaming off the weed seedlings before the carrots emerge.

We have tried both of these approaches, but in our experience there is no better weed control in carrot production than simply sowing the seeds into a thick layer of a mulching compost. We never lose beds to weed competition or spend time meticulously hand-weeding around rows of carrots. Moreover, the flame weeder feels like the most dangerous tool on our farm. On several occasions small fires have broken out in carbon-laden beds after flaming, and so we have stopped using our flame weeder. We also prefer to avoid stale seedbedding because it takes a bed out of production for too long. Compost mulching has made carrots a viable option for us.

SEEDER

We use an EarthWay seeder with the *light* carrot plate (purchased separately), or we use the Jang JP-1 seeder with the XY-24 roller. We adjust the brush on the JP-1 to allow for only one seed to drop at a time. Bolero, which is a big seed, needs the brush on the Jang (that determines the quantity of seeds that are dropped) to be higher, thus allowing for the size. But even still we have found that singulation with Bolero seed is difficult.

Sown too deep, carrot seed does not germinate well. Set the seeder depth at a quarter inch (6 mm) if sowing directly into soil, or at a third to one-half inch (8 to 13 mm) depth to sow in compost mulch.

SPACING

The ideal carrot spacing is 1 to 2 inches (3 to 5 cm) apart within the row. Distance between rows depends on the season and goals. Generally we sow five rows on a 30-inch (75 cm) bed; for baby carrots and Atlas we sow six rows. In the tunnels, however, we sow five rows of baby carrots per bed to increase airflow for the overwintered crop. Growers Zach Cannady and Kasey Crispin of Prema Farm in Nevada are excellent carrot producers. Their technique is to plant only four rows on a 30-inch (75 cm) bed, but plant the rows densely with carrots and achieve very high yields. Try a few different spacings and see what works best for you.

HARVEST AND YIELD

Check harvest readiness by pulling a few roots from different parts of the bed to evaluate size. At harvest, we bunch the carrots in the field. We keep a supply of rubber bands braceleted on our wrists, and we ease a band over the greens only twice—we do not wrap the bands more than twice in case a damaged carrot needs to be removed and replaced during washing.

Yields depend highly on goals, germination success, and degree of root maturity at harvest. We shoot for at least 200 bunches per 100-foot (30 m) bed of baby carrots, but often harvest 250 or more. The term *bunch* is a bit vague. Baby carrots may be a bunch of 8 to 10. Large carrots are 4 or 5 to a bunch. The larger the carrot, the more bunches you will get (often 400 bunches per bed),

Figure 9.2. Whenever possible, we sell carrots with the tops still on because the lively green tops denote freshness.

Figure 9.3. Marketing carrots is not a challenge. It is an easy crop to "pile high and watch fly" as the farmers market adage goes.

but that also means more time in the field. We have found at our market that larger carrots do not necessarily sell as well as smaller carrots. We rarely sell carrots by the pound, but when we do, we do not sell them with the tops on.

In the packing shed, we lay out carrots on a large metal mesh table, spray them down, and inspect for damage. We always haul in extra carrots unbunched for replacements—you can't easily spot damage in the field. We allow the roots to air dry (in the shade) for a few minutes before packing them into totes for market. We harvest no more than 2 days before market, because leaving the tops on for too long can drain moisture and sweetness from the roots, reducing their flavor and nutritional value. Any carrots intended for storage should have the greens removed at harvest.

INTERCROPS

Carrots are poor competitors when interplanted and grown beside another crop—greens are liable to grow tall but create small roots, for this crop. Instead, use carrots as a relay crop, sowing the carrots beneath mature crops about to be removed, or transplanting young plants of other crops into mature carrot stands. Carrots can also be added to beds primarily to allow another crop to benefit from the carrots' mycorrhizal associations. In this instance, harvesting a carrot crop would be a bonus. Pair carrots with cucurbits, nightshades, brassicas, legumes, alliums, spinach-family crops, and daisy-family plants.

MARKETING

For us, carrots are inarguably the easiest crop to sell. From June to August we can sell between 100 and 150 bunches of carrots per market, and many times we have sold over 400 bunches in a weekend during peak season. On the rare occasion that any carrots come home with us, we cut the greens off and store the roots for later sales.

Do not assume customers will balk at buying smaller carrots—test your market to find out. We sell large quantities of baby carrot bunches.

HARDEST SEASON: SUMMER

Germinating carrots in the heat can be tricky. The soil must remain moist and below 85°F (30°C) for carrot seed to properly germinate. Consistently irrigate the seedbed or add some kind of cover that will boost humidity—a tarp with the white side up or a heavyweight row cover. If you use a tarp, check underneath daily starting about 4 days after sowing the seedbed. Pull off the tarp immediately when you start to see signs of emergence.

Whenever you remove the tarp, do so in the later afternoon or at night to allow the young seedlings to adapt over the evening and next morning

Figure 9.4. These carrot seedlings are emerging through a light mulch covering of grass clippings that we applied just after sowing to reduce heat stress.

before being exposed to intense sunlight. The row cover, for its part, can cause soil to overheat. Make sure to keep the beds wet through either drip or overhead irrigation. Even after germination, the soil can get too hot and burn up the young sprouts, so beds must be kept fairly wet until established (an inch—3 cm—or so tall). Thirty-percent shade cloth suspended on hoops can be used for a week after germination to reduce heat stress, as well.

NOTABLE FAILURES

Perfecting our method for germinating carrots in compost mulch took a great deal of fine tuning. Our compost is loose and filled with half-rotten wood chips, which is great as a mulch but terrible for seed germination. At first, the carrot seeds were not staying moist enough. After we solved that problem, the seeds would germinate, but the young seedlings would perish in the beating sun of our Kentucky summers. More beds than I'd like to admit were lost due to poor protection after robust germination. Another notable failure was putting row cover directly on the soil of carrot beds in the spring. Though the presence of the row cover helps with germination, if the cover is not held off the ground on hoops, the wind whips the row cover up and down, which has the effect of cultivating the newly germinated seedlings right out of their beds. We also have concluded that row cover speeds up germination too much in the early spring. The carrots may grow too much before the spring frosts are over, and the exposure to cold may trigger premature bolting.

Arugula

Arugula is a popular salad green that we can sell year round to a wide audience. It is also a fast germinator that works well in a no-till system for the ease with which it can be terminated.

VARIETIES

We grow Astro year round. We've also had good success with Esmee in the spring.

ESTIMATED SEED QUANTITIES

When purchasing seed, we figure on 0.2 ounce (6 g) per 100-foot (30 m) row—six rows means 1.2 ounces (34 g) per 30-inch-wide (75 cm) bed. After I figure the total amount of seed we need, I usually buy an extra ounce (28 g) or two (57 g) just in case. If that reserve is still left at season's end, I generally mix the seed into the fall cover crop seeds (it winter-kills easily in our region).

BED PREP

Arugula seeds must have good soil contact and good moisture to germinate, but germination is quick compared to other crops—2 days when soil is 65 to 80°F (18 to 27°C). If sown in compost and the compost feels dry to the touch a few hours after a watering, it's a sign that it is not retaining enough moisture for consistent germination. In a deep compost mulch system, it is okay to compress the compost slightly to increase moisture retention—a weighted bed-roller is effective before or even after planting. If you do not work with a deep compost mulch system, keep in mind that arugula does not grow well in compacted native soils, and compaction should be addressed before sowing.

WEED CONTROL

As a 20- to 30-day crop, arugula outruns most weeds, but if you wish to harvest a second cutting, you must tend to the weeds first, or they'll contaminate the cut. Mulches help with weed control. A stale seedbedding prior to planting is effective, as well. Either way, make sure that you do not plant arugula into a weedy bed, and give the crop one cultivation before the canopy of arugula fills out. Also, remove any weeds by hand before harvest—it's easier to do in the field than when you are washing and packing.

PEST CONTROL

From fall to early spring, we do not have many issues with flea beetles, but in the summer we cover our arugula to reduce the incidence of the tiny holes characteristic of this pest.

Though most of our beds do not suffer from flea beetle pressure, generally speaking, we don't risk leaving the crop exposed for fear of losing a planting. Light row cover can protect arugula from most pests.

SEEDER

For baby arugula, we use the Jang JP-1 with the sprockets set at 9 in the back and 14 in the front. We use the YYJ-24 roller with the brush down. For years we employed the EarthWay with the cabbage-turnip plate to plant arugula with great success.

The four-row and six-row pinpoint seeders from Johnny's Selected Seeds are very effective, but not for sowing through a thick mulch of compost. The mulch tends to bunch up in front of the seeder—the four-row is worse than the six. We have transplanted arugula in the past, but because the density and number of plants needed, it can be very labor intensive to get a worthwhile stand unless selling "bunching arugula." The Paper Pot Transplanter from Paperpot Co is another option that we've tried with some success, but arugula is such an easy crop to direct seed in the tunnels almost year round in our climate that we don't find transplanting necessary.

SPACING

The goal is a thick stand of plants in the row, around a quarter to one-half inch (6 to 13 mm) apart. In the spring and summer we sow seven or eight rows per 30-inch (75 cm) bed. For the winter we sow only six rows per bed in the tunnel to allow for more airflow.

HARVEST AND YIELD

Speed and cleanliness are key components of successful arugula harvesting. If you plan to grow loads of arugula, the Quick-cut Greens Harvester from Farmers Friend is a must. This is an expensive tool, but it's a tool that cuts the job of harvesting down from half an hour to just a couple of minutes per 50-foot (15 m) bed. Just on the time savings alone, a few beds of arugula cut with the Quick-cut will cover its high cost. Using the Quick-cut, we always harvest arugula early before the heat of the day sets in. If the arugula is starting to bolt, but there are still good leaves, we harvest with a knife, working to avoid the bolting arugula.

We wash arugula gently and spin it dry in a salad spinner (that was converted from a washing machine). Bags are packed to the weight that we

Figure 9.5. We seed all of our arugula with the Jang. Here, I am sowing our fall arugula in September, a crop that we cut multiple times through the fall and winter.

Figure 9.6. A drill attached to the Quick-cut Harvester powers the serrated blades that do the cutting as well as drives the spinning green brushes that gently push the material into the green canvas catchment "basket." This cut and catchment system allow you to harvest several feet at a stretch before stopping to unload the basket.

need (for market that is 0.20 pounds [90 g]). Water always increases the risk of contamination, so when the leaves are clean at harvest we do not wash arugula afterward to avoid the risk of pathogens.

From the first harvest we expect about one pound (450 g) per bed foot (30-inch [75 cm] beds). The second harvest yields three quarters of a pound (340 g) or less per bed, and the yield continues to decrease with subsequent harvests. A tip I gleaned from Ben Hartman's book *The Lean Farm Guide to Growing Vegetables* is to lightly rake the arugula bed after harvest to remove any fallen leaves that can rot on the fresh growth and taint the next harvest (the book's cover shows Hartman doing this).

INTERCROPS

Although we do very little interplanting with arugula, we have found that it produces quite well in the tunnels below nightshades in the early plantings of April where row cover is not necessary yet. Arugula does work well in a relay crop, where sections in arugula beds are removed in the early spring to make space for long-season crops such as tomatoes (see the "Relay Cropping" section on page 187).

MARKETING

We pack arugula for market at 0.20 pounds (90 g) per bag. When we sell it in the marketing system, we use the one item for three dollars or two items for five dollars pricing. This works out to be between $12.50 and $15 per pound (450 g). If you want to avoid plastic bags, you can certainly do so. Grower Stephen Ciancioso of Buena Vista Gardens in Hawaii has put a great deal of work and thought into reducing his plastic usage. Stephen has used biodegradable NatureFlex bags for several years and recommends them. The bags are available from many online purveyors.

HARDEST SEASON: SUMMER

The challenge with summer arugula is twofold—the heat makes germination tougher and it also reduces the number of cuts because the plants tend to bolt earlier, thus affecting yield. Arugula may need to be misted several times a day on sunny days to cool it off. For those reasons, we prefer to grow our arugula in the tunnels all through the summer, because the plastic barrier provides a small amount of shade. In the field in the summer, we provide that shading by covering arugula with insect netting.

NOTABLE FAILURES

Summer is also the time when our arugula experiences the most flea beetle pressure. We must cover the arugula to protect it from the beetles. We used to use row cover, but even when we used the lightest-weight covers we would often lose part or all of the crop to rot or simply overheating. Investing in fine-mesh ProtekNet, albeit expensive, has made a huge difference. It protects the arugula without smothering it with trapped heat. Our hope is that in the future we'll reach a point when we do not need to cover arugula in the summer. We see this as a realistic goal, because we've observed that as the soil improves flea beetle populations go down. However, growing arugula without cover is a goal to approach slowly and with caution, using tools like ProtekNet as needed.

Garlic

Garlic is arguably the most important crop on our farm, both monetarily and in our kitchen. It sells well at market, goes in every CSA basket, and quite simply is an essential ingredient in most of our favorite meals.

VARIETIES

Music is our main garlic variety, and we've started trialing German White. Spanish Roja is another we've leaned on heavily. These are all hardneck varieties. The scapes produced by hardneck garlic are an additional marketable crop. We no longer plant any softneck garlic because the cloves in softneck bulbs tend to be too small. Those small cloves are harder for our customers to use in the kitchen. Storageability is longer for softneck varieties and they do better in the South than hardneck garlic. As the climate warms, we may need to reconsider growing softnecks.

ESTIMATED SEED QUANTITIES

In garlic production, cloves are the "seed garlic" (planting from true garlic seed is a rare and different thing). For Music, we estimate an average of seven useable cloves per bulb, and so we need about 85 to 90 bulbs per bed. We separate out small cloves and plant them in a separate row or two for green garlic production.

BED PREP

In the spring, we usually plant a mixture of sorghum sudangrass and summer legumes into beds slated for our garlic crop. In the midsummer, we terminate the cover crop by rolling and then covering the beds with a tarp. Beds are formed with additional mulching compost if needed after the tarp is removed. Instead of a tarp, we have also placed cardboard over the rolled cover crop and then formed beds with compost atop that cardboard. Planting garlic through the cardboard is a challenge; it's important to keep the beds very moist for several weeks to encourage decomposition and make punching through the cardboard easier. An auger bit on a drill can help to punch through thick cardboard.

WEED CONTROL

As a long-season crop, garlic requires long-term weed control. In our experience and climate, the deep compost mulch acts as both the perfect growing medium for garlic and the best weed control. Any stray weeds that appear can easily be cultivated out with a hoe. With hay or straw mulch, the garlic is slower to mature because of the cool nature of these mulches. Also, any weeds that pop up through the mulch have to be pulled by hand. Leaves are a great addition as a surface mulch as well, but they must be chopped for the garlic to poke through. Wood chips are often in short supply on our farm, but where possible we do spread wood chips in our paths between garlic beds because they provide excellent weed control.

POPPING CLOVES

Cloves are popped from their bulbs no more than two weeks before planting, so as not to risk rot or drying out. With some varieties, it's possible to smash the stem of the bulb against a hard surface to separate the cloves from the stem (there are many videos online that demonstrate this). Our variety of Music garlic cannot be handled that way. We strip the paper skin from the bulbs and slowly pop the cloves into a basket. We do this in October once the garlic has fully cured. We try to make this project a casual one, sipping beer in the evening and slowly popping garlic, because if we try to rush popping six or seven thousand cloves in a couple days, our hands end up hopelessly sore.

SPACING

To mark four rows we use the Gridder, which is a simple tool for marking rows and spacing made by Neversink Farm Tools. We space our garlic 7.5 inches (19 cm) apart in the row. Cloves are pressed with the germ side down. The germ is the brown flat edge closest to the roots on the bulb—it's what you usually cut off when cooking. We plant at a depth of roughly 4 inches (10 cm) below the compost surface in our climate—generally the germ goes slightly into the native soil. Colder climates may require a deeper planting to avoid heaving from repeated freeze and thaw cycles, which can hoist garlic cloves upward toward the surface. Plant any cloves intended for green garlic as deep as 6 inches (15 cm) to increase blanching of the stems, or hill them up in the spring. We like to plant garlic in mid- to late October.

HARVEST

For green garlic, we start harvesting as early as we can—usually mid-March. Dense rows are intentionally planted for green garlic, but we also thin out any plants that are lagging behind the others in size.

The next crop chronologically is garlic scapes. We pull these flowering stalks of the hardneck varieties absolutely as early as we notice them forming so as to increase the scapes' tenderness and to reduce any chance of reduced bulb weight. Removing the scapes as soon as possible (by hand) can increase the garlic bulb yield by 20 to 30 percent, according to research from the University of Guelph in Ontario.[1] Plus scapes are tastier at that stage anyway. In the field, we bunch scapes with rubber bands in groups of 10 to 12.

Following scapes is the coveted fresh garlic. This is garlic that has begun to bulb up. The cloves are differentiated, but the papery skins between cloves have not yet formed so the entire bulb is edible—garlic at its peak, to my mind.

Finally, once the last garlic tops have browned out between 50 and 60 percent, we pull the entire crop. This freshly harvested garlic is still considered fresh garlic for two weeks, and it bruises easily if mishandled. We tell helpers to treat the garlic like eggs.

Figure 9.7. Green garlic is the stage before the plants bulb up. We sell it from March until scapes begin to elongate in May. That's when fresh garlic begins—this fresh garlic bulb is cleaned up and ready for market.

Figure 9.8. Most scapes should be pulled green and early, but the scapes on elephant garlic make for an excellent cut flower and attract loads of native insects.

Figure 9.9. When you're farming with kids, sometimes you run out of time to pop scapes while the crop is still in the ground! That's okay, leaving scapes attached does not ruin the bulbs.

If too many roots and soil are coming up with the garlic bulbs (usually when soil is fairly moist), we pull the garlic up to the surface by its stem and cut the roots off with a knife, leaving them in place. This slows down the harvest but leaves the soil and roots in place. The entire plant is brought back to the barn. Tops are removed at about two inches (5 cm) above the bulb, and those tops are composted. We like doing this project in the barn because garlic harvest is at the end of June or early July when the weather can be painfully hot. Bulbs are then sorted as seed, market, CSA, or kitchen (the damaged bulbs), and laid out on drying racks with a fan set to blow air lightly across them. We stack our racks to consolidate them, which conserves space. After two weeks we reposition the fan to be less direct. The bulbs stay on the racks—notably *not* cleaned, which increases storage life—until needed for market or kitchen. A dehumidifier may be necessary in more humid environments.

INTERCROPS

Garlic is a poor competitor, but it is too valuable a crop to mess up. If you wish to try intercropping with garlic, I recommend a cautious approach. Once the garlic has matured, interplant another crop right before harvest as a relay crop. Or plant garlic into a crop that will not survive the winter, such as chard or lettuce. This can be done by drawing a string across the bed for guidance, and using a trowel or drill with an auger bit to conduct the planting. This will of course be very time consuming, but it's a great use of space if you're working on small acreage.

MARKETING

We sell green garlic at the market (3 or 4 in a bunch) and garlic scapes (10 to 12 per bunch) just like we do for other crops: one bunch for three dollars, or two bunches for five dollars. We sell bulbs of fresh garlic and cured garlic for three dollars each, and we do not apologize for it—garlic is valuable and customers rarely complain (especially once they taste it). Our bulbs are also substantial in size (approximately three inches (8 cm) in diameter), which is important. We would not sell a small-diameter bulb for that price. By late August, when the bulbs begin to dry and shrink, we sell two bulbs as a unit.

HARDEST SEASON: SPRING

Hardneck garlic has a vernalization period. That means the garlic cloves must be exposed to a cold period to reliably bulb up in the summer following planting. The vernalization period is at least 40 days below 40°F (4°C), though they do not have to be continuous. If the developing plants do not receive that period of cold, they will grow large, but individual cloves will

Figure 9.10. Austrian winter peas coming up underneath garlic. It seemed like a successful pairing, but garlic yields were dramatically compromised due to too much competition between crops.

not form. Softneck garlics do not need as much vernalization and therefore tend to be the preference of growers in USDA Hardiness Zones warmer than Zone 6b. You can also use lighter-colored mulches such as straw and hay to keep the soil cool longer. Studies on artificial vernalization (refrigeration) of garlic bulbs before planting show mixed results. If planting garlic late or in Zones 7a and warmer, I suggest sticking with a softneck like Nootka Rose. Cloves planted in the spring can produce a good crop of green garlic, but they will not make substantial bulbs.

NOTABLE FAILURES

We learned the interplanting lesson the hardest way possible—by thinking it was a success. In 2018, we sowed Austrian winter peas into our garlic after planting time. In the spring, the field was absolutely stunning with the mix of healthy-looking peas and garlic plants. I bragged that it was our tallest garlic ever. This was, as it turns out, true but also not a good thing.

Garlic is a poor competitor, and during its critical period of competition—a concept I was not aware of at the time—it was challenged by the peas. The competition forced the garlic to put more energy into its leaves than its roots. The bulbs were smaller, but I assumed this was because we had not pulled the Austrian pea plants early enough. So the next year we did a small sample trial with half of a bed of garlic planted with Austrian winter peas again, and the rest of the garlic patch without any peas. We pulled the peas when they reached 18 inches (45 cm) tall to use as a mulch. The bulbs from this small trial were the smallest in the patch. Now, it is possible the interplanted garlic was more nutrient dense, but it would require a dedicated field trial and analysis to determine that. For our purposes, the significant effect of the intercropping was to shrink our garlic yield. The lesson here is that if you intend to intercrop with garlic, give the garlic plenty of space. Also, choose a shorter crop than peas as the competition.

Lettuce

Lettuce is a crop that is in high demand all year, but it has a number of small needs that must all be met to have consistent success in production.

VARIETIES

For summer heads, we grow Starfighter, Muir, Cherokee, and Grazion. For spring and fall heads we prefer Skyphos, Muir, and Cegolaine. For winter heads, we love Muir, our year-round workhorse. For cut lettuce—also called spring mix or lettuce mix—we grow the Salanova Foundation Collection mix all year.

ESTIMATED SEED QUANTITIES

At the time of this writing, we are not direct seeding lettuce at our farm. All production is with transplants. All of our seed quantities are based on heads per bed, even for lettuce mix. We account for 640 heads per 100-foot (30 m) bed, or 160 heads per row. On occasion, we grow romaine as large heads; we space these one foot (30 cm) apart—so 100 per row, or 400 per bed. Overly large heads do not work in our marketing system of a set price per head. We prefer seeds pelletized with NOP-compliant materials for seeding lettuce, as the naked seeds are painfully small.

Although we no longer direct seed lettuce in the garden, we always enjoyed using Johnny's Selected Seeds Encore Lettuce Mix for that job. This mix makes a great leaf lettuce and we sowed it at the recommended rate of one ounce (28 g) for every 400 feet (122 m) of row. However, we prefer the Salanova and other head lettuces for making our own lettuce mix because of the flavor that develops over the long maturation in the field and the reliability of transplanting versus direct seeding.

BED PREP

Lettuce is a shallow-rooted, fast-growing crop that is very forgiving in terms of bed prep. Generally, we check for compaction and address it if necessary with a light broadforking. Next we fertilize the bed if we're not following a

Figure 9.11. Muir is winter and summer hardy. We often sow it with Salanova in case we need a little extra cut lettuce. When we don't, we use the Muir lettuce as a head lettuce.

cover crop, and we replace any areas that need mulch with compost. Rarely do we plant lettuce into a terminated cover-cropped bed because we reserve cover crops for heavier feeders and longer season crops.

PEST CONTROL

Our biggest issues with lettuce pests are not bugs eating the lettuce, but bugs hanging out inside the heads. This includes spiders, occasionally slugs, cucumber beetles, crickets, earthworms (yeah), and others. We have learned that the best way to avoid passing along these hidden bugs to our customers is to bubble the lettuce in water using a homemade Shop-Vac bubbler—a pipe with holes in it attached to the blower on a Shop-Vac. After a 1-minute bubbling treatment, we allow the heads to soak, fully submerged, for 10 minutes. At that point the vast majority of stowaways crawl out of the heads and can be removed with a towel or by hand. Or they float to the surface to be skimmed off.

In the winter, *Sclerotinia* drop (also known as lettuce drop) can be an issue. This fungal disease can absolutely ravage a stand of lettuce. Fungal diseases are best avoided by supplying ample airflow and reducing leaf moisture—allow for 10 inches (25 cm) between plants when transplanting. If row covers are needed at night, remove them during the day if possible. We also open up tunnels whenever outside temperatures are above freezing.

WEED CONTROL

Transplanting instead of direct seeding is the best form of weed control for lettuce. Beds should be clear of weeds before planting, because when you put a young lettuce transplant into ground that does not have any weeds, it will have several week's head start on competition. The canopy fills out rapidly, blocking out any young weeds. In the peak of summer, we start harvesting lettuce within three to four weeks after transplant; there is rarely time for weeds to grow and compete.

GERMINATION

We grow our lettuce in soil blocks made with a Stand-up 35 Soil Blocker sold by Johnny's Selected Seeds. This maker puts 105 blocks on a tray, and six trays supply enough to plant a 100-foot (30 m) bed (with a few extra that we can use for gaps or replacements). Summer lettuce is our more important crop, but germination can be tricky because lettuce seed goes dormant above 85°F (30°C). We cope with this by keeping lettuce seed in the cooler. We then sow blocks in the evening and leave them in the shade of our shed until we see sprouting (usually within 2 days during the summer). Once sprouts emerge, we promptly move the trays to the greenhouse. If temps are

Figure 9.12. Shade cloth can help reduce stress for summer lettuces as they establish in hot conditions. We leave the shade cloth in place for two weeks after transplant.

not predicted to drop below 85°F (30°C) overnight on sowing days, we set newly sown trays in the cooler overnight, remove them in the morning, and place them in the shade until germinated.

Many growers use the Paper Pot Transplanter from Paperpot Co for lettuce, but it does not work in our system because the mulch is too dry and the chains of paper pots used in the Paperpot system do not break down easily and wind up blowing all over the farm (at the writing of this, the manufacturers are working on hemp chains that will more easily decompose). Transplanting by hand is slower, but we have found that a single person working efficiently and using both hands at once can plant an entire 100-foot (30 m) bed in 30 to 40 minutes.

SPACING

We space lettuce heads 7.5 inches (19 cm) apart in the row, with rows also 7.5 inches (19 cm) apart (four rows per bed). The exceptions are winter lettuce, which we plant at about 10 inches by 10 inches (25 cm by 25 cm) to allow for better airflow. Other exceptions are large heads such as romaine for restaurants or special orders. We plant them 12 inches (30 cm) apart (three rows per bed) with 12 inches (30 cm) between plants. Occasionally we grow mini-romaine heads like Truchas, and we space those closer to 6 inches (15 cm) apart but still in four rows, 7.5 inches (19 cm) apart.

HARVEST AND YIELD

During the summer, lettuce harvest takes place first thing in the morning before the heat sets in. In the coldest parts of winter, we wait until after the sun is up and the lettuce has thawed.

For spring, fall, and winter crops we can get multiple cuttings of Salanova for lettuce mix—typically, the first cut yields 0.20 to 0.25 pounds (90 g to 115 g) per head, or 1.2 to 1.5 pounds (544 g to 680 g) per bed foot. It decreases with each subsequent cutting. The Quick-cut Harvester is not effective for harvesting this crop because the Incised and Sweet Crisp Salanova varieties have a core. It's important to avoid cutting this core if regrowth is desired.

We harvest leaf lettuce into bins or totes. Lettuce leaves are bubbled as described previously, spun, and then packed into bags at 0.25 pounds (115 g) per bag. We also sell a mixture of lettuce and pea shoots (which we call the micro mix): We add 1 pound (450 g) of pea shoots to every 5 pounds (2 kg) of lettuce, and bag that mixture at 0.20 pounds (90 g) per bag—a smart idea suggested by our 2018 interns Adam and Amanda Dilley. We sell these bags using the one for three dollars, two for five dollars model.

For head lettuce, we cut the heads at the base and remove the entire plant, always leaving the roots in the soil. We remove any bad leaves and

Figure 9.13. These winter lettuces were planted in early October. The goal with winter lettuce is for the heads to reach this size before December when the days are short and photosynthesis slows down.

then pack the heads into bins in the field. In the packing shed, we clean the heads by bubbling them as described previously, spin them dry, and put them into clean bins or coolers for market.

INTERCROPS

Lettuce is our intercropping champion and we plant it with practically everything. Also during the main season we always sow extra trays of lettuce heads to fill gaps in beds of carrots, green onions, radishes, beets, or turnips. Lettuce pairs extremely well below nightshades and cucurbits before they have filled out their beds. Broccoli and long-season brassicas also pair well with head lettuces.

MARKETING

Displaying lettuce in plastic bags is difficult because the moisture condenses and fogs up the bags, making them less attractive to customers. We work around this in a few ways. First, we keep only as much bagged lettuce on the table as we can reasonably sell in the subsequent half hour. To make smaller piles look more bountiful, we shake each bag and trap the air inside with a twist tie added right before displaying it. We set out a large sign (see figure 9.14) to inform customers that we have lettuce available, even if it's too hot

Figure 9.14. Having a bountiful market table is important, but we use this sign that Hannah made as a stand-in for displaying bagged lettuce when hot conditions could lead to rapid wilting and ruin the product.

or sunny to pile bags on the table. Once a bag shows condensation, we take off the tie and put the bag back in the cooler.

With head lettuce we make bountiful piles, but display only as much as we can sell before it wilts. We mist heads lightly throughout the day and switch them out with others from the cooler to reduce any wilting. Displaying any amount of wilted greens on your table make the entire table look less fresh. We never display bags or greens directly in the sun, as the sun makes miniature greenhouses out of the bags—trapping heat and creating condensation—or rapidly wilts the loose greens.

We sell head lettuce as one item for $3 or two for $5. Overall, our cut lettuce earns us $10 to $12 per pound (450 g). The pea shoot micro mix earns $12.50 to $15 per pound (450 g). And earnings from the small head lettuces vary from $6 to $15 per pound (450 g), or $2.50 to $3 each.

HARDEST SEASON: SUMMER

The work needed to market local fresh lettuce from June through September feels overwhelming at times, and there is a reason why—lettuce is a labor of love to grow in the heat. For summer lettuce to be successful in our climate, every detail must be carefully attended to. Germination temperatures must be monitored. The greenhouse where transplants grow cannot be too hot. The plants should be hardened off for a few days in the direct sun before transplanting. Transplanting itself should be done quickly into soil that is not hot (irrigation can help cool a bed quickly) then the plants must be thoroughly watered in. Lettuce should be misted multiple times per sunny day with overhead irrigation throughout its growth. Thirty percent shade cloth for the first two weeks after transplanting helps reduce heat stress, bolting, and loss. Lastly, lettuce has to be harvested early in the morning to avoid bitterness.

Ray Tyler of Rose Creek Farms is a masterful summer lettuce grower, and he direct seeds lettuce all summer. His trick is that all of his beds are under caterpillar tunnels, and he keeps them very moist from seeding to germination. My friend Elam of Flat Creek Farm in eastern Kentucky is one of the only other summer lettuce growers at our market. He says he successfully transplants lettuce into beds covered with white plastic during the growing season with drip tape underneath the plastic. Again, there are many ways to grow summer lettuce, as long as each step is done correctly.

NOTABLE FAILURES

We have had no shortage of lettuce failures on our farm: In our work developing a summer lettuce program, we direct seeded lettuce into soil that was too hot and got no germination. We have since learned that we prefer to

Figure 9.15. Lettuce is a great intercrop, but we went overboard here trying to combine four rows of head lettuce with a row of field tomatoes. The lettuce planted at the edges of the bed did fine, as they generally do; the lettuce heads planted in the two center rows were smothered out as the canopy of tomato foliage developed and became dense.

transplant our lettuce rather than direct seed for the quality and consistency of the product.

Once we finally got germination down, we failed to adequately cool the crop as it grew. We finally realized that using a small amount of shade cloth in the first two weeks, followed by some light overhead irrigation to mist the lettuce, was enough to keep it from bolting or perishing in the heat.

Not suspending the row cover high enough above the lettuce during a freeze results in leaves that get "burnt" from the cold and so they become unsaleable. We have since started adding taller hoops and pulling our row covers taut.

One big close call we had was not washing the lettuce well enough and giving a customer a head that still had a slug tucked inside. (We are thankful they found it before it ended up on someone's fork!) Fastidiously monitoring for bugs and properly soaking the lettuce heads has helped us to eliminate this potential disaster.

Sweet Potatoes

A crop that we always make ample room for in our garden plan is heirloom sweet potatoes. This crop does extraordinarily well in our hot summers without irrigation, and it provides food for ourselves and our customers all winter.

VARIETY

We grow only one variety, the Golden Nugget sweet potato. It's a locally adapted heirloom that performs well in our climate. However, there are an amazing array of sweet potatoes out there. What's more, they can be grown as far north as southern Canada. Maybe even farther. Choose shorter days-to-maturity varieties if you are in a cooler climate than USDA Hardiness Zone 6a.

ESTIMATED SLIP QUANTITY

If you purchase slips, figure on 150 to 160 slips per 100-foot by 30-inch (30 m by 75 cm) bed. We grow our own slips, and we use the rule of thumb that one medium size, half pound (230 g) potato equals 8 slips. We save about 150 sweet potatoes (about 75 pounds (34 kg)) from our fall harvest to use for generating slips. This provides enough slips to plant eight beds.

GROWING SLIPS

There are many ways to grow slips. In the past we dug a trench in the soil and placed whole sweet potatoes in a single layer in the trench. We covered the potatoes with a few inches (8 cm) of creek sand (not commercial bagged sand) and allowed them to sprout. We have not had access to creek sand

Figure 9.16. These sweet potato slips in moist, mulched soil will quickly recover from the shock of being planted outdoors and begin their growth.

lately, so we simply place a single layer of the sweet potatoes right on the bottom of the deep harvest bins and cover them with our mulching compost. The slips grow through the mulch or sand and we gently pull them off the potatoes. The process is described in more detail in the following sections.

BED PREP

The looser the bed at planting time, the less soil disturbance will be necessary during harvest. If you have access to straw and ample amounts of compost, build lasagna beds—they are perfect for sweet potatoes due to their inherent friability. In our region where straw is not available (or reliably chemical-free) we plant into deep compost. In our rotation, sweet potatoes follow early spinach. The early spinach is mowed to the ground in May. We then broadfork if need be, and then tarp the beds for two weeks. After removing the tarp, we spread five pounds (2 kg) of alfalfa meal per bed followed by compost. We limit the compost to a single strip a few inches (8 cm) deep down the center of the beds, aiming to conserve the compost. We do this because sweet potatoes are a less-profitable crop for us, so we aim to minimize inputs. After planting and a single cultivation a couple weeks after planting, we then fill the space between the strips of compost with peat moss as a light mulch for moisture retention. We use the 24-inch (60 cm) Landzie peat moss spreader for this task. One compressed bale of peat moss covers around 400 square feet (37 sq m).

PEST CONTROL

The worst pests for sweet potatoes are deer aboveground, and voles or gophers belowground. We are fortunate not to have issues with below-ground damage on our sweet potatoes, but deer can eat the tender greens all the way down to the dirt. If there are deer in your area and your gardens are not near your house (where the presence of dogs may be a deterrent), deer fencing is essential.

As for disease, we occasionally see a small amount of the fungal disease called scurf on our sweet potatoes, but it has largely disappeared since we began growing no-till. Good drainage and healthier soils decrease occurrences. Any roots with scurf are hard to market. If scurf becomes overly abundant, a new seed stock may be necessary.

WEED CONTROL

Sweet potatoes vine out and create a substantial canopy, but during the establishment phase, they must be cultivated as often as you see emerging weeds. Once the canopy fills in, an occasional hand weeding is necessary to prevent any weeds present from going to seed; it is okay to step on the sweet potato vines while weeding. In fact, in some regions and cultures, vines are even lifted out of the pathways and flipped over onto the beds to increase

Figure 9.17. Scurf is a fungal disease caused by *Monilochaetes infuscans*. It does not harm yield or flavor, but it does make the potatoes less marketable.

root size. We do not move the vines, but we do harvest sweet potato greens from our pathways (described in the following sections), and we have not seen a significant reduction in yield since we started this practice.

SPACING AND PLANTING

We plant our sweet potatoes after the absolute last chance of frost (late May to early June). The day before planting, we start a batch of compost tea. The compost tea is then mixed with compost and soil (if available) to essentially create mud for "mudding in" the sweet potato slips. Each slip is carefully popped from its sweet potato and the roots are dipped into the mud to coat them with the nutrient- and biology-rich slurry. This helps them establish. If a slip does not have any roots it will not survive. A slip does not have to have an abundance of leaves. In fact, fewer leaves with more roots is our preference, because such slips tend to be less susceptible to sun damage.

Using a piece of rebar or something equally sturdy, make holes eight inches (20 cm) deep at eight inches (20 cm) apart in the row. Slide a slip gently into each hole and press soil up around the stem. Only a small amount of the plant needs to be above ground. The plants should then be watered in thoroughly. We prefer one row per 30-inch (75 cm) bed, though if we were working on slightly wider beds we would plant two rows per bed, staggering the slips in each row.

HARVEST AND YIELD

Sweet potatoes continue to grow as long as the plant is still alive (up to a frost). In order to avoid being surprised by football-sized roots, we start checking our potatoes at the beginning of September. To harvest, we first mow the vines using a flail mower. We set the mower a little high so as not to risk damaging potatoes. Then we find the first cluster of roots and gently pop it with a fork or shovel. The looser the soil, the easier the harvest. So long as the soil is not too wet, we can lift the whole cluster of potatoes out on the vine, and then gently grub for any left behind. In the past we used a potato plow attachment on the walk-behind tractor. That technique disturbed the soil heavily, and possibly created deep compaction, too. It also damaged a fair number of the sweet potatoes. It is a better tool for harvesting regular potatoes, and I do not know of a plow attachment for walk-behinds that is designed for harvesting sweet potatoes. The extra labor in hand-popping the potatoes has an upside—there are significantly fewer damaged potatoes, and every pound (450 g) damaged represents a loss of two to three dollars. Handle sweet potatoes gently, even though most damage heals during the curing process. (The damaged potatoes may not be marketable, but they will be edible.) We expect rows at the edge of a sweet potato patch to yield around 225 pounds (102 kg) per 100-foot (30 m) bed, and closer to 200 pounds (91 kg) for inside beds.

CURING

To reach their full flavor potential and sweetness, sweet potatoes must be cured for two weeks following harvest. Curing does not have be a complicated procedure to produce a great crop. Sweet potatoes should be kept at 80 to 85°F (27 to 30°C) and 100 percent humidity for 14 days. To accomplish this we have used our greenhouse (wide open) in the late fall. To create the humidity, we place thick blankets over the sweet potatoes to trap the moisture they are naturally releasing. If you live in an arid environment, add some amount of humidity by placing bowls of water underneath the blankets alongside the potatoes, or run a humidifier.

After the curing process, they can be stored in dark areas, in baskets, anywhere from 55 to 70°F (13 to 21°C). We often store sweet potatoes for slip

Figure 9.18. We cure our sweet potatoes in a small greenhouse. Notice that soil remains on the surfaces of the sweet potatoes, which helps to protect them from damage as they cure. We keep the potatoes covered with a blanket throughout the curing process.

production and eating ourselves in cool areas of our house all the way until spring. The sweet potatoes remain edible until early April when they begin to convert those sugars back into starch in preparation for slip production.

INTERCROPS

Because sweet potatoes "vine out" low to the ground, there is a lot of potential for intercropping either in the sweet potato beds or by planting beds of other crops between the beds of sweet potatoes (which also helps increase sweet potato yield, as the rows that yield most are always the rows without sweet potatoes on both sides). Any intercrops just have to be able to grow above the canopy. Long-season crops like dent corn or sorghum are perfect, as they need no midseason tending. Tall varieties of bush beans work, as well. I have heard of growers using okra between their sweet potatoes. Sweet potatoes perform well in a small amount of shade and neither crop needs much fertility, so this pairing has much potential to boost the revenue of sweet potato plots.

MARKETING

Sweet potatoes are an essential crop for our family for our winter meals, but they are not a wildly profitable crop. We set three or four small sweet potatoes in a berry quart container for $3 or two for $5. On average, that earns us $2 to $3 per pound (450 g) at market, so a perfect bed that yields 200 marketable pounds (91 kg) of sweet potatoes earns us $400 to $600. To boost that revenue potential we are working to increase the market for sweet potato greens. Though not well known in the United States, sweet potato greens are a widely consumed crop throughout the world from South America to South Africa. The greens are cut from the pathways and at the edges of the sweet potato plots by lifting the vine and stripping off the most tender shoots. The vines are then bunched in rubber bands in the field. We have always given these to the CSA in August to break up the summer monotony of tomato-family crops, and we are slowly working to make them a market staple.

NOTABLE FAILURES

Sweet potatoes are an easy crop to grow once established, so our biggest failures have all arisen from curing and storage mishaps. Curing sweet potatoes in a greenhouse is a fine tactic, so long as the temperatures do not get too hot. One year we left the greenhouse closed for a sunny day by accident and hundreds of pounds of potatoes literally cooked in the heat and then rotted. We also had a heater fail in our storage room on a very cold night, freezing the roots and resulting in a massive loss of sweet potatoes in the middle of winter. The sweet potato is a forgiving crop, but it has its limits.

Beets

Both a cold- and heat-hardy crop, beets are one my favorite year-round vegetables to grow. Beets also make an excellent intercrop. And when piled on the table, they really draw customers in, and we can sell nearly every bunch of beets we bring.

VARIETIES

We grow Boro, Zeppo, Chioggia, and Touchstone Gold beets.

ESTIMATED SEED QUANTITIES

We transplant most of our beets; we purchase 1,200 seeds per 100-foot (30 m) bed or 300 seeds per single row. We do not direct seed beets often because (like Swiss chard) beet seeds are multigerm seeds that produce multiple plants from one seed, which results in plantings that are too dense. For times when we do plan to direct seed, we purchase approximately 1,600 seeds per bed.

BED PREP

Beds are fertilized and a small amount of compost mulch is added if needed. We soak the trays of beet seedlings in compost tea or extract before transplanting. During the bulk of the summer months—late May through late August—we plant all of our beets under 30 percent shade or in the tunnel. The small amount of shade helps establish the beets but also produces more marketable greens. For direct seeding, beets take 5 to 7 days for germination and prefer soil temperatures around 85°F (30°C). Keep the soil moist until germs appear.

PEST CONTROL

So long as we plant beets under some amount of shade, we do not have issues with *Cercospora* leaf spot or the other common leaf diseases. Blister beetles occasionally attack our beets. If we spot blister beetles on other crops (usually the spring chard when it begins to weaken over the summer) we cover the beets with insect netting such as ProtekNet or a lightweight row cover at transplant or upon emergence.

WEED CONTROL

As with other crops, transplanting the beets into a mulched, weed-free bed is our strategy for weed control. However, if direct seeding, the story is a little different. Beds are prepped with ample compost mulch. To speed up germination we will add either heavyweight row cover or a tarp with the white side down for a few days—you want this process to take the least amount of time possible to discourage competition. Beets are a 50- to 70-day crop, post emergence, so we cultivate at least once after they have fully emerged, even if we don't see weeds.

SEED STARTING AND SEEDING

We sow all of our beets into 1⅛-inch (29 mm) soil blocks made with the Stand-up 35 Soil Blocker. We place only one seed per block. In the spring, we place soil blocks on a heat mat in our greenhouse to speed germination. In the summer, they can germinate in a well-ventilated greenhouse with no bottom heat required. Blocks are covered lightly with vermiculite or soil mix to retain water and encourage germination. To make sure that mice do not have access to these blocks, we place them on a table that mice cannot climb up onto (metal legs usually work). Mice will devour beet seeds.

Beet seeds vary in size, so when sowing beet seeds directly into beds, we pair the size of the beet seed to the seeder or seed plate. Generally, we use the LJ-24 roller on the Jang with the sprockets set at 14 in the back and 9 in the front to space the seeds about 2.5 inches (6 cm) apart. The beet plate on the EarthWay seeder is effective as well, though the chances of sowing the

Figure 9.19. Every full bed of beets is harvested twice. We first go through and harvest the bigger beets, then we allow the smaller beets to size up for one more week before harvesting the rest of the bed.

crop too thickly are higher with the EarthWay because it lacks the accuracy of the Jang. If direct seeding, we like to start a couple of trays of extra beets to fill in any gaps.

SPACING AND PLANTING

When transplanting, we set four rows about 7.5 inches (19 cm) apart, with the beets transplanted at around 4 inches (10 cm) apart in the row. A rainy day is a good opportunity to thin the blocks in the greenhouse to two seedlings per block. This is more efficient than trying to do the thinning in the field. For direct seeding we sow five rows of beets, with the seeds spaced two to three inches (5 to 8 cm) apart. We thin direct-seeded beets only if necessary.

HARVEST AND YIELD

We harvest beets after all the tender greens are out but before crops such as carrots are ready. When beets are largely two inches (5 cm) in diameter, we bunch them with rubber bands in the field, peeling off any yellow or rough leaves. Three large or four medium-sized beets make a marketable bunch.

Figure 9.20. We harvest and bunch our beets in the field, then take them to the wash area for inspection and to remove any dead leaves that we missed. The ability to sell large quantities of beets depends heavily on them having clean roots and healthy-looking greens. Taking a few extra seconds to clean up the bunches makes all the difference.

We also bunch up any remaining small beets in bunches of five or six to sell as "baby beets"—a popular item at our market. Beets are then gently washed on our washing table, drip-dried, or gently shaken off. We pack them in totes and store them in the walk-in for market. Harvest never occurs more than 2 days before market. Our goal is 300 bunches per 100-foot (30 m) bed, which works out to an earnings between $700 and $900 per bed.

INTERCROPS

Beets are fairly competitive, and they are one of my favorite intercrops. They also do not make mycorrhizal associations, so we find it benefits them to have other crops around that do. My favorite pairing is with solanaceous crops such as peppers, tomatoes, and potatoes. They also work well in a relay cropping scenario where, for instance, the bigger beets are thinned from a bed in the first harvest. A round of summer squash is then planted into that bed. While the squash establishes, we complete the beet harvest the following week. Beets, like head lettuce, are a great filler among poorly germinated carrots, as well.

MARKETING

We sell beet bunches with greens attached at three dollars for one bunch or five dollars for two. We pile the beets high on the table—their color tends to draw in customers extremely well. We mist the greens throughout the day.

HARDEST SEASON: SUMMER

Transplanted beets can have a difficult time establishing in the heat of the sun. We like to transplant them early in the day on a cloudy day or in the late afternoon. We also water them in very well and mist them a few times during hot days to cool the plants down. Beets are more heat tolerant than lettuce, but they still require occasional cooling in our climate. Planting them below the shade of a tunnel or shade cloth helps with establishment and growth.

NOTABLE FAILURES

Our biggest struggles with beets have come from overseeding them, resulting in an overly thick stand that will not produce sizable beets. Thinning beets is extremely time consuming! On the flip side of that coin, when direct-seeding beets into compost, we have found that poor germination happens. Our solution has been to choose the seeder based on the size of a particular batch of seeds, as mentioned previously. It also helps to cover the beds and keep the beds extremely moist for germination. Lastly, mice have contributed to no shortage of beet failures in our greenhouse, and we have learned to germinate the beets on something rodent-proof. One technique is to put them on a table with metal legs that mice cannot easily climb (card tables work well for this), and place the table away from a wall. A slick plastic tote or barrel can work as well, too, where the beet trays are simply piled on one another until germs appear.

Cherry Tomatoes

In our market it is hard to compete with the established growers on larger slicing and heirloom tomatoes, so we have filled the cherry tomato niche. These are extremely fun crops to grow because with the long season you get a chance to really steward the plant into success through pruning and fertilizing.

VARIETIES

Sun Gold, Black Cherry, Bumble Bee, and Supersweet 100 are our main cherry tomato varieties. We also really like Blush and Mountain Magic, among other specialty or saladette types.

ESTIMATED SEED QUANTITIES

We purchase 1.1 seeds per plant that we intend to grow. The extra 10 percent is insurance to offset the possibility of poor germination.

BED PREP

We plant cherry tomatoes very early in the season, so making sure the soil is warm, healthy, and free of weeds is the goal. Where possible, we like to have a winter-kill cover crop precede the cherry tomatoes to create the best possible fertility and soil structure. This is easier to accomplish in the field than in a greenhouse where the winter growing space is supremely valuable. There are options for relay cropping cherry tomatoes into arugula or carrots. When a tomato planting has not been preceded by a cover crop, we amend the bed with our usual amendments before planting. We broadfork if needed, and we often put down a sheet of landscape fabric two weeks prior to planting to warm the soil. The landscape fabric can be left in place or removed. I prefer to remove it for more interplanting potential and ease of adding amendments to the soil.

After planting, we watch for signs of nutrient deficiencies or excesses, foliar diseases, and pest outbreaks throughout the season. We then amend the plants as needed in situ through foliar sprays derived from Korean Natural Farming techniques and soil amendments. When there is no landscape fabric, we add soil amendments using the EarthWay with the pea plate, pouring the amendments into the hopper. The notches on the pea plate distribute them slowly.

Figure 9.21. There are many ways to grow cherry tomatoes. Here they are in a single leader system and planted through landscape fabric for early production.

PEST CONTROL

Our biggest pest challenges with cherry tomato production are aphids and hornworms. Aphids can be a sign of excess nitrogen, which should be addressed by consulting with an agronomist. The best way to combat these two pest challenges is prophylactically. We purchase ladybugs and lacewings early in the season while the tomatoes are still in trays to preemptively stop aphid outbreaks. There are some environmental concerns related to importing ladybugs. Many ladybugs sold commercially are wild harvested while dormant, and they may transmit disease or parasites to your local populations. Lacewings are a little less concerning in that regard, but the goal should always be to build up native populations of predators by not using pesticides, providing habitat, and planting plenty of flowers, so that importing beneficials becomes less necessary. Our hornworm strategy is to plant flowers nearby, such as sweet alyssum, cilantro, and dill. Their flowers help attract parasitic wasps that lay their eggs in hornworms.

Grafting tomatoes is increasingly common as a measure to prevent disease problems. The rootstock varieties used are resistant to specific soilborne diseases. We have not yet explored grafting, though many growers have had success and seen increased yields with this process. Grafting may be worth trialing on your farm, especially if you intend to grow tomatoes in the same place year after year (a common practice in tunnel production on small acreage).

WEED CONTROL

Being a long-season crop, cherry tomatoes require a thick mulch for the best weed suppression. Any mulch will work, though I recommend being very conscious of the effect of mulch on soil temperature. Studies suggest the best root growth for tomatoes occurs right around 70°F (21°C)—this is why many growers plant their early tomatoes into plastic or landscape fabric, which helps warm the soil with its dark color. We have used landscape fabric for early plantings to great effect, but we prefer to use compost mulch because its dark color helps warm the soil below and the nutrients in the compost also feed the soil. In the field, we use hay or straw mulch through the summer to cool the soil and keep it closer to that 70°F (21°C) range.

SEED STARTING

Tomato seeds require warm soil to germinate. We sow seeds in two-inch (5 cm) soil blocks and place the trays of blocks on a heat mat to germinate. This takes about 4 days at 75 to 80°F (24 to 27°C). Some growers use mini soil blocks to start the seeds, but we find that step unnecessary. Instead we prefer to add a few extra seeds to each tray of blocks. That way, extra seedlings will sprout in some blocks. We can simply pull these out if needed to fill in any empty blocks where seeds failed to germinate. Once the plants are four to six inches (10 to 15 cm) in height, we form balls of soil mix around each in lieu of using the four-inch (10 cm) soil block maker. The advantage is speed—we can make two of these balls in the time it takes to make one four-inch (10 cm) soil block. (We do this with pepper transplants, as well.)

SPACING AND PLANTING

Of all the spacing and training methods we have tried for cherry tomatoes, what we like best is to set plants 18 inches (45 cm) apart in a single row on a 30-inch (75 cm) bed. We transplant before the plants are flowering when they are at about 12 inches (30 cm) in height. We train them to a double leader as described in the following section. This allows us to interplant other crops alongside the tomatoes. Alternatively, two rows 15 inches (38 cm) apart can be planted on a 30-inch (75 cm) bed, with plants 12 to 18 inches (30 to 45 cm) apart and trained to a single leader, also described in the following section. In the field our spacing is closer to 30 inches (75 cm) apart in the row.

When the soil is ready, we use an auger bit on the end of a portable drill to open up a planting hole about six to eight inches (15 to 20 cm) deep, depending on transplant size. We fill the hole with compost tea or extract, which helps to establish good biology at the roots. We remove the lower leaves of the tomato plants (so as not to bury the leaves) and then place the root ball in the hole and refill the hole. We lightly remulch the area around the plant in case weed seeds were brought to the surface in the digging process. If grafting, do not heap up soil above the graft because this will allow the scion variety to form roots, defeating the purpose of the grafting.

TRELLISING, TRAINING, AND PRUNING

Plants, fruit, and leaves must remain off the ground to maintain good airflow, which reduces development of disease. This is where a good trellis comes into play.

The number of ways to trellis tomatoes is nearly infinite. It is an incredibly important subject for market gardeners. I'm going to describe the two methods that we use most at Rough Draft Farmstead. If intensive tomato production is something you're particularly interested in, check out the resources section at the back of this book for listings of videos and books to reference.

The first trellis we use is a wire trellis. This is a strong wire or cable that hangs above tomato plants in the tunnels at a height of roughly nine feet (3 m). (Some growers create structures that can support wire trellises in the fields, as well). This trellis wire has to be very strong because the plants

Figure 9.22. Special clips help to hold up tomato plants in a wire trellis system. The clips go around the tomato stem and gently snap onto the string. Note that this clip is holding this plant upright, but any clips that will bear weight for the plant must be placed directly under a substantial branch. The clips are available at any horticultural supply center.

become extremely heavy when loaded with fruit. We fix this wire to the structure of the tunnel itself to ensure that it is well supported. The plants are tied to twine from special spools that hang from this wire; this keeps the plants upright.

The reels contain roughly 50 feet (15 m) of strong twine that can be unspooled as a plant grows. Once the plants reach the wire, the reel is flipped over or string is unwound. This action, which helps keep the plant tops from hitting the ceiling of the greenhouse, is repeated weekly throughout the season as the vines continue to grow. However, if you only unspooled more and more string, the plants would end up laying on the ground, which you do not want. The goal is to keep the plants—and more specifically, the fruit—off the ground. So after a little line is let out of the spool, each spool is then moved laterally along the support wire, working systematically down the row. This is called "lowering and leaning," and it is a standard practice in horticulture. It has the effect of keeping the plant off the ground and giving each plant more space to grow. In essence, rather than simply growing vertically, the vines end up growing horizontally, as shown in figure 9.25. This method of lowering and leaning also works well with cucumbers to maximize space and keep the plants from hitting the high tunnel plastic.

For field production, the trellis is simpler. We place eight-foot-tall (240 cm) T-posts at every third plant, directly in between the plants. We usually do this a couple of weeks after transplanting so that we can cultivate once if needed before there are posts in the way. String is then woven between the plants in the Florida weave pattern, as shown in figure 9.23. A new string is added once a week. Outdoor hanging trellises can also be created on which to support tomatoes, though they must be extremely stable to contend with heavy winds and rain. Josh Sattin of Sattin Hill Farm in North Carolina has a clever trellis system detailed on his YouTube channel, which he describes in the video "A Trellis to Make You Jealous." It is essentially half-inch (15 mm) electrical conduit pipes extending horizontally in midair. This trellis is created by threading the conduit through 1¼-inch (32 mm) PVC tees that are inserted over the tops of seven-foot-tall (215 cm) T-posts. Strings are then tied to the conduit and dropped down to the plants.

Training and pruning tomato plants is an art. The myriad details are largely beyond the scope of this book, but the basics are easy enough to explain. Double leader and single leader are the two main ways to train tomato plants. The number of leaders refers to the number of stems. So a single leader tomato is just a plant with one stem. A double leader is a plant with two main stems. Hannah and I have tried many training and pruning methods, and we generally prefer to train our cherry tomatoes in the tunnel to a double leader. If you are new to training and trellising tomatoes, however, start with a single leader.

Tomatoes put out small shoots known as suckers at the junction between a leaf stem and the plant's main stem. If allowed to grow, each sucker will

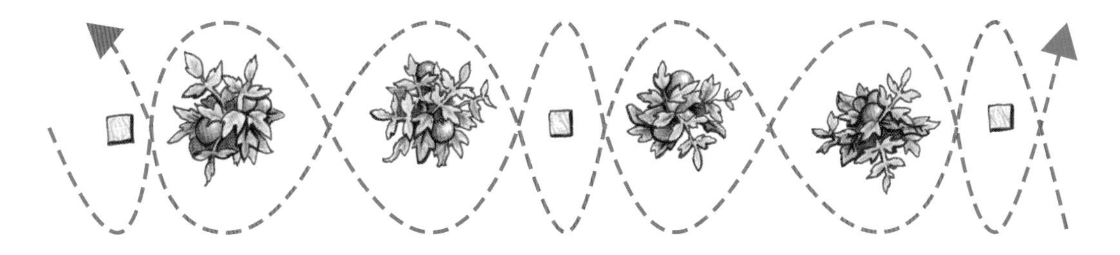

Figure 9.23. A fast trellis in the field is made by going around the plants with biodegradable twine in this weaving pattern called the Florida weave.

Figure 9.24. Suckers emerge between the main stem and the base of a leaf stem. These will keep growing, becoming fruit-producing branches unless removed.

become another stem that produces leaves and fruit. To make a double leader, we allow the first sucker that forms on a plant to grow. We train the main stem and the sucker stem each onto a string that's attached to the supports. The results resemble a letter Y. Each branch of the Y is trained up its own line and is lowered and leaned in opposite directions, as depicted in figure 9.25. This essentially doubles your tomato production per plant. It also gives the roots a bit more room at the soil surface, because there is only a single row of tomatoes as opposed to two rows, as is common with single leader systems. This allows for more interplanting, too. That said, single leader systems can be highly productive, as well, and we often use a single leader system for later-season plantings of tomatoes.

Pruning is critical for airflow. Every week we remove all new suckers and maintain only 13 or 14 leaves on each main stem. All the other leaves are removed as well as any empty fruit stems once all the fruit has been harvested. This intensive pruning persists until the late summer when the plants slow their rate of growth (with the decrease in sunlight). We cease pruning in September and allow the plants to be a little wild, which saves us the time we would have to devote to pruning. In the field we prune off branches that are touching the soil to discourage disease. Other than that we let the plants grow as they please. This is why we space them so widely in the field, to accommodate all the top growth.

HARVESTING AND YIELD

During the peak of the season, we harvest at least twice per week, looking for tomatoes that are at least 85 percent ripe. Our preferred harvest bins have wide bottoms, and we stack the fruit no more than a few inches (8 cm) deep to avoid crushing the lower level of fruit. The bins are then placed in the cooler. Counter to popular understanding, cold temperatures do not ruin tomatoes, but they do slow the ripening. As for yield, each plant will be a little different. Because we sell them by half pints (240 ml) and not pounds, it's not

Figure 9.25. As tomato plants near the top of their trellis, string is unspooled, then re-hooked to the top wire a few inches farther along, which allows the vines more lateral growing space while keeping the tomato fruits off the ground, hence: lowered and leaned.

always easy to translate (a half pint is roughly a half pound (230 g), but that's not exact). We shoot for an average of at least one pint (500 ml) per week per stem. So over the course of a 10-week period in which they are most productive, a single plant can earn roughly $100 dollars, and our plants usually produce well beyond that 10-week period. The first plants go into the ground in late March or early April, and their production usually lasts from late May until late October, though we have harvested into November in the tunnels.

INTERCROPS

Tomatoes are an excellent crop for first experiments with intercropping because they are highly competitive. With our preferred planting scheme of a single row of tomatoes down the middle of the bed, there is room on either side to plant a fast-growing crop while the tomatoes mature. Radishes, lettuce, beets, and green onions all work well in that space. In between tomato plants, sweet alyssum can also be planted to attract pollinators and beneficials.

BED FLIP

One of the biggest question marks in tunnel tomato production is when to remove the tomatoes and replant the area with winter crops. There is no easy way to make this decision, but I have a simple suggestion: Winter crops will succeed only if they are planted by a certain cutoff date, and that date varies by region. In central Kentucky, that date is October 10. As long as we plant winter crops by this date, the plants will have enough time to mature before the cold, low-light months of December and January when plants cease to grow (winter growing is really just winter harvesting). Tomatoes must be removed by that day to allow for a successful winter crop to follow in that bed. This typically works out well—the tomato plants have stopped producing much at that point anyway. We have been experimenting with planting winter leeks into existing tomato beds as a relay crop so that there is no need to remove the tomatoes until after a freeze, though that is still an idea in its infancy as I write this book.

MARKETING

We mix varieties together and sell them in half pints (240 ml) as one item for three dollars or two for five dollars. Customers appreciate this size, as a pint box can hold a large number of cherry tomatoes—too many for someone who simply wants to make a salad.

HARDEST SEASON: EARLY SPRING

Many factors are working against you in the early spring right around planting time. While young tomato plants are still in the soil blocks, they can be highly susceptible to aphids. Using a professionally made soil mix will help mitigate this issue. We also preemptively employ ladybugs and lacewings in March inside the greenhouse to immediately quash any aphid outbreaks.

We often plant our first tomatoes in unheated greenhouses around April 1—a few weeks before our last chance of frost. Cold soil temperatures can greatly stunt tomato plants. Using compost mulch helps warm the soil, as does landscape fabric. The other issue with the early planting is that some spring cold snaps can be severe. In fact, much of the East Coast suffered a late freeze in May 2020. At that point our tomato plants were large and already hanging from the trellis. To protect them, we had to take them off the trellis and double cover them with row cover, as most tomatoes are not hardy below 34°F (1°C). If you can affordably heat your tunnel, this will make a significant difference in their production and your stress level.

NOTABLE FAILURES

Finding the balance between too much fruit (generative growth) and too many leaves (vegetative growth) has been a significant challenge. If the balance is too far off in one way or the other, there can be a long gap in production. The excessive presence of lush leaves is a sign of too much nitrogen. Overfruiting is a sign of too little nitrogen, too much stress, or dramatic changes in temperature. Maintaining this leaf-fruit balance is also one of the fun things about growing tomatoes—unlike many of the crops described here, the grower can "steer" the success of this crop over the course of the season by adding amendments based on how the crop is performing. In that way, tomatoes are one crop that can be corrected if something goes wrong. To really get the most out of cherry tomatoes, it would be worth working with an agronomist and doing a series of sap and soil analyses throughout the season to manage their output and health. Several growers I've talked with have recommended working with the agronomists at Advancing Eco Agriculture, headed by John Kempf. Kempf's blog and podcast are also excellent general resources for managing soil health. (For more information, see "Resources and Recommended Reading" on page 255.)

Some Thoughts in Closing

Although these seven crops are the mainstays of our business, they have not always been so and, honestly, they may not be in the future. After we transitioned to no-till and got to know our market better, these crops floated to the top. However, as markets change, as our skills as growers improve, as new techniques come to light, we may gravitate away from these crops and toward others that make more sense.

In fact, while nearing the completion of this book, Hannah and I decided to move farms. It was a positive move—one that we'd been contemplating for some time. We loved our farm, but with its long, narrow, winding driveway, it was not feasible for us to do any significant amount of on-farm sales. Our new farm, however, has a building near a main road we are currently fashioning into a farm store. And a rural farm store will likely require a different array of crops than what we typically grow for the farmers market. We may have to improve our skills at growing slicing tomatoes and heirlooms. We may have to grow more cucumbers and find clever ways to produce sweet corn profitably. But we don't dread that challenge, we look forward to it. We have faith in our growing systems. We have an ever-expanding understanding of plant and soil biology. Our path to success in this new location is to listen to our customers and do what humans are built to do—steward the soil.

APPENDIX A

Cover Crop Use and Termination Guide

The following listings provide guidance on when and why to seed a particular cover crop, what to pair it with, and how to terminate it. I do not include seeding rates for individual crops because that information is contextual. Start with the recommended seeding rate per acre provided by the purveyor. If you are planting a mix of crops, divide the recommended rate by the number of crops in the mix. From there, consider other factors that have an influence on germination and crop growth, such as climate, time of planting, whether you're seeding or broadcasting, soil type, seed size, and weather conditions. Then adjust the seeding rate up or down. This is a matter of judgment, and you'll learn from your experiences and mistakes. Take good notes about seeding rate decisions and observations of how crops fared.

BROADLEAF PLANTS

Buckwheat (*Gaopyrum esculentum*)

Season: Plant in late spring or all summer. Prefers warm soils; very slow to germinate and grow in cold soils.

Mix Recommendations: Hard to mix with other cover crops as it outcompetes most of them and also goes to seed faster. Consider sowing after other cover crops have emerged if you intend to mix.

Termination: After plants have been in flower for a few days, mow the crop low on the stems. Easy to winter-kill.

Benefits: An excellent phosphorus and microbial gatherer as well as a weed suppressant. It is quick to flower, attracting beneficials—such as predatory insects, pollinators—and honeybees. Great for a quick cover between spring and fall plantings here in the South.

Cilantro (*Coriandrum sativum*)

Season: Plant in fall for fall and winter cover crop, or in spring for beneficials.

Mix Recommendations: Most fall and winter cover crops. Sow into fall covers and cut as marketable item.

Termination: Winter-kills where frost and freeze cycles are frequent, but can overwinter in milder climates or climates with heavy snow. Crimping at flowering doesn't always kill cilantro; mowing can be effective. Tarp or solarize for best results.

Benefits: Effective fall and winter ground cover. Flowers are great for beneficials. Has potential for phytoremediation of lead- and arsenic-contaminated soils.

Lacy Phacelia (*Phacelia tanacetifolia*)

Season: Sow in early spring for flowers or in late spring for cover (the crop does not

flower without vernalization). Can have low rates of germination—sow liberally.
Mix Recommendations: Spring cover crops including oats, wheat, barley, and field peas.
Termination: Mow low or crimp after flowering. Tarp or solarize.
Benefits: Excellent soil conditioner; provides tilth. Flowers are very attractive to beneficials. Can absorb excess nitrogen and calcium in soils.

Lettuce (*Lactuca sativa*)

Season: Sow in early spring or fall.
Mix Recommendations: All fall and winter cover crops.
Termination: Winter-kills easily. Can also be crimped or mowed very low at flowering.
Benefits: Grows well with a variety of other crops; fills in a canopy well. Bolts fast in heat but is slow to mature seed, which can be good for pollinators. We use it in place of oats with field peas.

GRASSES AND GRAINS

Annual Flax (*Linum usitatissimum*)

Season: Plant in spring and summer.
Mix Recommendations: Peas, oats, or with most summer cover crops. Very small seed and does not create substantial biomass—best in a mix.
Termination: Can winter-kill but generally requires a hard frost, similar to peas.
Benefits: Makes mycorrhizal associations. Relatively fast growing. Can reach flowering stage in 50 days and continue flowering for two weeks.

Barley (*Hordeum vulgare*)

Season: Plant in late fall or early spring. Prefers cool, dry conditions.
Mix Recommendations: Field peas and other tender legumes as well as cereal rye.
Termination: Not as hardy as cereal rye. Winter-kills in regions with several freeze and thaw cycles or when temperatures drop below 17°F (-8°C) without snow pack.
Benefits: Nutrient scavenger, especially of nitrogen. Thick root systems provide biomass and can increase soil friability. Fast grower and allelopathic; effective for outcompeting and reducing weed populations. Can be grown in colder regions than most other grains can tolerate.

Cereal Rye (*Secale cereale*)

Season: Sow in late summer or late winter. Great for sowing into mature cash crops in the fall.
Mix Recommendations: Crimson clover, vetch, Austrian winter peas, and most other winter cover crops.
Termination: Can be crimped at milk stage. Otherwise, cut or mow and then solarize or tarp.
Benefits: Extremely winter-hardy cover crop with allelopathic properties that inhibit weed growth (particularly grasses). Voluminous biomass above and below ground.

Japanese Millet (*Echinochloa esculenta*)

Season: Sow in late spring or all summer.
Mix Recommendations: Cowpeas and other summer legumes and warm-season grasses or grains.
Termination: Mow low after flowering. Winter-kills when sown late in the season.
Benefits: Excellent summer cover crop that can grow 4 feet (120 cm) tall in 45 days. Creates large root systems; proficient at soil building. Nitrogen scavenger.

Oats (*Avena sativa*)

Season: Plant in late summer for a crop that will winter-kill, or plant in late winter for a spring cover crop.
Mix Recommendations: Field peas for winter-killed cover crop or with field peas and lacy phacelia in the spring.
Termination: Winter-kills with several freeze and thaw cycles, forming a light mulch. Crimp or mow plants in milk stage then tarp or solarize.

Benefits: Nutrient scavenger and retainer. Creates substantial biomass and softens soil. Weed suppressant.

Pearl Millet (*Pennisetum americanum*)

Season: Sow in late spring or all summer.

Mix Recommendations: Japanese millet, cowpeas, and sorghum sudangrass

Termination: Mow or scythe high for regrowth or mow low at flowering to terminate.

Benefits: Grows well in dry areas with low soil fertility. High protein content; a rich source of nitrogen for microbes to convert to plant-available forms.

Sorghum Sudangrass (*Sorghum × drummondii*)

Season: Sow in late spring or all summer.

Mix Recommendations: Soybeans, millet, cowpeas, and sunn hemp.

Termination: Not easy to kill. Mow or crimp then tarp for several weeks in the summer. Winter-kills very well (do not graze after frost because this crop creates prussic acid, which can be toxic to ruminants).

Benefits: Disease, weed, and nematode suppressant. Compaction reliever and soil improver. Grows fast and tall—supplies biomass and provides great fall mulch for long-season fall crops. Mow or graze multiple times to 6 to 8 inches (15 to 20 cm) in height to encourage regrowth and soil building.

Sunflower (*Helianthus annuus*)

Season: Sow in late spring or all summer.

Mix Recommendations: Summer cover crops such as sunn hemp, cowpeas, soybeans, and sorghum sudangrass.

Termination: Winter-kills easily. Crimp or mow very low after flowering. Root balls are substantial and take a few months to break down.

Benefits: Nitrogen scavenger. Drought tolerant and easy to terminate. Flowers are used medicinally by beneficials—bee populations that feed on sunflowers show lower rates of infections and disease.

Teff Grass (*Eragrostis tef*)

Season: Sow after last frost; this is a warm-season grain. Seeds can be small, so this crop can be hard to establish evenly.

Mix Recommendations: Other summer grains including sorghum sudangrass, sesbania, and sunn hemp.

Termination: Can be crimped; a subsequent tarping or solarization in addition is most effective. Winter-kills easily.

Benefits: Fast growing, good for weed suppression. Does not flower easily, and thus is low maintenance. Can survive a light mowing; works well with sorghum sudangrass and as a forage crop.

Triticale (*Triticum aestivum × triticosacale*)

Season: Sow in fall or late winter. Triticale is a cross between rye and wheat.

Mix Recommendations: Rye, vetch, Austrian winter pea, and crimson clover over the winter. In the late winter with wheat, peas, and oats.

Termination: Treat like rye, or mow and cover with tarp or clear plastic for several weeks. Does not winter-kill easily.

Benefits: Nutrient scavenger. Builds soil and produces a long-lasting mulch.

Wheat (*Triticum aestivum*)

Season: Sow in late summer or late winter.

Mix Recommendations: Cereal rye in the late summer, or with oats, peas, and lacy phacelia in the spring.

Termination: Crimp or mow at milk stage then tarp or solarize.

Benefits: Nutrient gatherer, weed suppressant, and soil builder. Mitigates erosion.

LEGUMES

Austrian Winter Pea (*Pisum sativum*)

Season: Sow in late summer or late winter.

Mix Recommendations: Winter-hardy grain crops in the fall; or oats, phacelia, and wheat in the later winter.

Termination: Does not winter-kill easily. Can be crimped or mowed. Tarp or solarize for several days afterward to ensure termination.
Benefits: Winter hardy, nitrogen fixer. Produces more substantial aboveground biomass than other varieties of peas.

Berseem Clover (*Trifolium alexandrinum*)
Season: Plant in early spring or late summer.
Mix Recommendations: Winter-hardy annuals or spring annuals such as oats and peas.
Termination: Winter-hardy in warmer regions; can regrow after mowing. Crimp or mow in the spring and follow up with long occultation or solarization.
Benefits: Large biomass creator. Excellent weed suppressant and nitrogen fixer. Can be used as a living mulch in northern climates as long as it is mowed regularly.

Cowpea (*Vigna unguiculata*)
Season: Sow in late spring or all summer.
Mix Recommendations: Summer cover crops such as sorghum sudangrass and sunn hemp. Interseed with corn.
Termination: Mow or crimp then tarp in late summer. Winter-kills easily.
Benefits: Fast-growing summer legume that fixes nitrogen and surpasses weeds. Drought tolerant. Exceptional for beneficials due to extrafloral nectaries, which are glands located on the stem that secrete nectar and attract pollinators.

Crimson Clover (*Trifolium incarnatum*)
Season: Plant in late summer or late winter.
Mix Recommendations: Rye, wheat, oats, barley, vetch, phacelia, and Austrian winter pea.
Termination: Crimps well with rye; is not tolerant of low mowing. Very winter hardy.
Benefits: Scavenges and fixes nitrogen. Attracts many types of beneficials when flowering, including pirate bugs, which feed on thrips and other pests.

Field Pea (*Pisum sativum*)
Season: Plant in late winter or late summer.
Mix Recommendations: Oats are the most common companion, but also wheat, phacelia, lettuce, and barley. Also any summer cover crops for winter-killing.
Termination: Survives light frost but winter-kills after a few freezes. Can also be mowed low at flowering. Crimp then tarp or solarize when growing in a mix.
Benefits: Nitrogen fixer and weed suppressant. Edible flowers and shoots for an early market crop. Also excellent as a winter-killed cover crop when planted in late summer.

Mung Bean (*Vigna radiata*)
Season: Sow in spring and summer after last chance of frost. Prefers warm soil.
Mix Recommendations: Other summer legumes such as cowpeas and sunn hemp. Can generally keep up with buckwheat. Good with sorghum sudangrass to fill in gaps in the canopy.
Termination: Terminates easily with a low mowing at flowering. Can also be tarped or solarized.
Benefits: Generally affordable seed. Low lying and easy to terminate. Very fast grower and can produce beans in 60 days.

Red Clover (*Trifolium pratense*)
Season: Sow in spring or fall in USDA Hardiness Zones 4 and warmer. Oversow into maturing cash crops in the late summer for winter cover crop. There are two types of red clover: medium and mammoth.
Mix Recommendations: Oats, vetch, rye, and wheat. Can also be broadcast into existing cash crops to provide ground cover, or as a nurse crop.
Termination: Medium red clover is hard to terminate and winter hardy. Mow or crimp then tarp or solarize for several warm weeks. Mammoth red clover is easier to terminate.
Benefits: Nitrogen fixer and soil builder. Attracts beneficials. Regrows after mowing, so has potential as a living mulch for certain crops and in pathways.

Sesbania (*Sesbania exaltata*)

Season: Sow in spring or summer. This is a tropical plant that does not grow well where summer nights are cool.

Mix Recommendations: Sorghum sudangrass, millets, and other tall warm-season cover crops.

Termination: Winter-kills easily. Otherwise, mow or crimp then tarp.

Benefits: Nitrogen fixer. Attracts parasitic wasps, lady bugs, and other beneficials. Can handle high mowing. Grows fast. Drought tolerant.

Soybeans (*Glycine max*)

Season: Plant in early summer as summer cover or in late summer for a winter-killed crop.

Mix Recommendations: All summer covers including sunn hemp, sesbania, and sorghum sudangrass. If planting with buckwheat, sow the soybeans about one week before the buckwheat or presoak the soybean seeds overnight.

Termination: Winter-kills easily or mow low to the ground after pods appear.

Benefits: Nitrogen fixer, fast growing but slow to go to seed. Fills in canopies nicely. The seed can be expensive but it is easy to collect and save your own supply of seeds for replanting.

Sunn Hemp (*Crotalaria juncea*)

Season: Sow in late spring or all summer.

Mix Recommendations: Summer cover crops such as sorghum sudangrass and sesbania.

Termination: Mow then tarp or solarize. Winter-kills easily.

Benefits: Ideal summer cover crop. Fast, grows tall, and creates a great deal of biomass and mulch. Heat and drought tolerant.

Sweet Clover (*Melilotus officinalis, M. albus*)

Season: Sow in spring or summer. Needs vernalization for flowering.

Mix Recommendations: Winter-hardy crops like rye, vetch, and Austrian winter pea. Seed into beds mid- to late-season where you would like a cover crop the next year.

Termination: Difficult to terminate due to deep taproot. Mow or crimp then tarp for several weeks. Because clover is difficult to terminate and control, do not let plants go to seed.

Benefits: A superb nutrient miner with a deep taproot. Excellent for beneficials in second season. The flowers are a great nectar source for honey production.

Vetch, Common (*Vicia sativa*)

Season: Sow in late summer or late winter; moderately winter-hardy.

Mix Recommendation: Mixes well with winter-killed cover crops in colder climates, and winter-hardy cover crops in warmer climates

Termination: Winter-kills in colder climates. Otherwise the deep taproot makes it slightly difficult to terminate. Wait until it flowers if possible, then crimp or mow and tarp or solarize for several weeks.

Benefits: Nitrogen fixer and flowers are great for pollinators. Effective weed suppressant and is slightly allelopathic. Creates substantial biomass.

Vetch, Hairy (*Vicia villosa*)

Season: Sow all season; more winter hardy than common vetch.

Mix Recommendations: Other winter hardy grains and grasses such as wheat, rye, and barley. Also mixes well with Austrian winter pea and crimson clover.

Termination: Deep taproot makes vetch slightly difficult to terminate. Wait until it flowers if possible, then crimp or mow and tarp or solarize for several weeks.

Benefits: Superb nitrogen fixer and phosphorus scavenger. Produces substantial, sprawling biomass. Good habitat and flower source for beneficials.

BRASSICAS

Alyssum (*Lobularia maritima*)

Season: Plant in spring or summer. Not winter hardy.

Mix Recommendations: Almost any long-season crop. Great for undersowing among tomatoes and peppers.
Termination: Goes to seed easily, but can be mowed or lightly cultivated out. Winter-kills with a few hard frosts.
Benefits: Low-lying annual flower that attracts beneficials, especially parasitoid wasps. Not much biomass but good insectary addition to summer cover crops.

Mustard (*Guillenia flavescens*)

Season: Sow in fall for overwintering.
Mix Recommendations: Rye and crimson clover.
Termination: Mow after flowering but before pods form. Tarp or solarize to ensure termination.
Benefits: Biofumigant for several soilborne pathogens. Manages nutrients well. Excellent weed suppressant, and good for beneficials when allowed to flower early in the spring.

Tillage Radish (*Raphanus sativus*)

Season: Sow in late summer, one month from first frost dates. Can also be sown in early spring.
Mix Recommendations: Most legumes. Effective in a winter-killed cover crop such as lettuce, peas, and oats. Also mixes well with rye, vetch, and other overwintered crops.
Termination: Winter-kills in climates with regular freeze and thaw cycles. Otherwise, mow low or crimp (which will snap the roots standing above the surface) and allow for the roots to decompose for several weeks.
Benefits: Compaction reliever. Scavenges nitrogen in the fall and stores it over the winter for the succeeding crops.

Turnips (*Brassica rapa*)

Season: Sow in early spring or late summer.
Mix Recommendations: Tillage radishes, rye, vetch, crimson clover; or wheat, oats, lettuce, and field peas.
Termination: Fairly winter hardy but winter-kills with several freeze and thaw cycles. Otherwise, mow at flowering and allow roots to break down before planting a subsequent crop (tarps speed up this process).
Benefits: Can retain nitrogen and other minerals over the winter. Relieves some compaction. We often sow turnips as part of a mix of winter cover crops, but we harvest many of the roots for market as the canopy fills out.

APPENDIX B

Critical Period of Competition and Interplant Pairings

The following listings detail the critical periods of competition for many common market vegetables, along with recommendations of favorable interplanting combinations. I have included a few crops, such as arugula, that do not have defined critical periods from the research. For those crops, I give guidance based on my experience growing them and on what I know of their growth habits.

ARUGULA/ROCKET

Critical Period: Arugula is fast growing and very competitive. Though there is no established critical period of weed control for arugula, beds must remain weed free throughout their life to ensure healthy growth and efficient harvests.

Pairings: Arugula is easy to terminate, so it is a nice crop to sneak in below taller crops like tomatoes, sunflowers, peas, and corn. Cut out plants at soil level when harvests are done. Also effective as the first in a relay cropping sequence.

BASIL

Critical Period: Basil should be kept weed free for the first four weeks after transplant (eight weeks after emergence). However, it can compete with many types of companions as long as airflow is sufficient so as to avoid fungal diseases.

Pairings: Studies have shown improved yields when pairing basil with corn. Tall sweet peppers and heavily pruned tomatoes work, too. Basil does nicely as a relay into lettuce or beets. Basil flowers make an excellent insectary, as well.

BEANS

Critical Period: Beans generally compete well against other crops. Yield loss from competition happens before flowering, though it is also fairly small at about 3 percent.[1]

Pairings: Carrots work well with beans if carrots are mature when beans are transplanted. Brassicas and tomato-family crops also coexist nicely with beans. When grown with corn and squash, pole beans are part of the indigenous Three Sisters technique of interplanting; beans pair well with both of those crop families.

BEETS

Critical Period: Beets are a strong competitor. Research suggests that beets should be competition-free for two to three weeks post-emergence to avoid a 1 to 5 percent loss.[2] You can eclipse this period by growing transplants, so avoid direct seeding when interplanting.

Pairings: Transplanted beets make a great understory for all nightshades (including potatoes). Beets do not make mycorrhizal associations, so pair them with green onions, *mature* carrots, parsley, ginger, daisy family, and nightshades. Beets are also excellent in a relay sequence with summer squash.

BROCCOLI

Critical Period: Broccoli is a decent competitor though ideally it should be allowed to grow free of competition for roughly two weeks after transplanting.[3]

Pairings: Broccoli is nonmycorrhizal, so it pairs well with mycorrhizal crops like lettuce and green onions. Parsley, cilantro, and celery are also good companions where water and space allow. Broccoli pairs well with some taller grass crops such as corn (if broccoli is planted on the northern, shadier side of the corn) to help with cooling in summer. Nasturtiums, buckwheat, or any flowering insect-attracting plants help attract parasitoids that help control cabbage worms. It is easy to overseed a bed of broccoli with parsley or cilantro.

CABBAGE

Critical Period: Fairly competitive. After transplant, beds should be kept competition-free for three weeks to avoid yield reductions. Competing crops should not be left in the beds for more than five weeks.

Pairings: Brassicas benefit from interplants that make mycorrhizal associations. Head lettuce and green onions are excellent pairings, or transplant cabbages into mature carrots. Nasturtiums or any flowering insect attractant help attract parasitoids. Cilantro also makes a friendly ground cover below cabbage in the fall.

CARROTS

Critical Period: Carrots do not thrive in competition. Not only do they suffer in the presence of even small amounts of weeds, but many common weeds carry root-knot nematodes that can attack carrots, and many others may host aster yellows disease. Interplanting carrots should almost exclusively be done when they are mature or when the preceding crop is nearly ready to be removed.[4]

Pairings: Mature carrots excel in a relay cropping sequence, or when they have a lot of room. We sometimes undersow summer squash beds to carrots before the squash take over a bed. Our most common approach, though, is to plant squash, head lettuce, beans, onions, beets, or tomatoes into already-mature carrots (within two weeks of harvest).

CAULIFLOWER

Critical Period: Cauliflower can be particularly susceptible to competition in the first two weeks after transplant; in one two-year study weeds in that period dropped the average weight of cauliflower heads by 41 percent.[5]

Pairings: Cauliflower is nonmycorrhizal so it benefits from the presence of lettuce or onions. Nasturtiums, buckwheat, or any flowering insect attractant help increase the presence of parasitoids. Cilantro and parsley make a superb ground cover for cauliflower in the fall.

CELERY

Critical Period: There is not an established critical competition period for celery. It is a slow-germinating and slow-growing crop, so avoid adding any competition to celery beds that could shade out the celery plants too aggressively.

Pairings: Short-season crops such as spring radishes, head lettuce, and beets have all worked for us. One study demonstrated how leeks interplanted with celery improved weed suppression without compromising yields.[6] We were surprised by a successful pairing with peppers in a wet area of our garden where the celery seemingly helped sop up excess water, and both crops succeeded. I believe this pairing would be better on beds wider than 30 inches (75 cm) to allow for a little more light to reach the celery.

CORN

Critical Period: Corn is the perhaps the most well studied in terms of critical periods. As with all crops, its critical periods vary based on climate and variety, but in general, keeping corn competition-free for the first 50 days from seeding, or until six independent leaves are visible, is ideal for maximizing yields.[7]

Pairings: Corn and legumes tend to work well together; cucurbit-family crops also pair well with corn. Upright corn stalks can serve as support for vines and also supply shade for tender summer crops like lettuce and arugula. Fast-growing crops such as radishes or turnips also work with corn.

CUCUMBERS

Critical Period: Beds should remain competition-free for up to four weeks after emergence (whether in soil or cells), and interplanted crops should be removed from the bed within five weeks to avoid yield losses.[8]

Pairings: Radishes or transplanted head lettuce and beets all make excellent pairings. Dill, nasturtiums, and smaller members of the daisy family, such as marigolds, also have potential. Green (young) garlic has been shown to have numerous beneficial effects when planted with cucumbers.[9]

EGGPLANT

Critical Period: Eggplant does not generally compete as well as other solanaceous crops. Avoid much competition for the first two months after transplant.

Pairings: A fast round of radishes, green onions, arugula, beets, or lettuce are safe bets. Adding some flowers such as sweet alyssum, nasturtiums, and marigolds can help deter pests.

FAVA BEANS

Critical Period: A fairly competitive crop, fava beans still prefer to have no competition in the first 30 days after emergence.[10]

Pairings: Fava beans are a versatile crop for pairings, though it depends on where you live. Plant favas in the fall in temperate, warmer climates (USDA Hardiness Zones 7b and warmer), therefore undersowing fall crops such as kale that don't overwinter can be an option. For cooler climates, it is best to undersow the fava beans right before they are harvested for a relay cropping sequence. Potatoes, celery, parsley, parsnips, and green onions are all good candidates.

FENNEL

Critical Period: Fennel does best with 40 to 50 competition-free days after emergence.[11]

Pairings: Fennel is notoriously antagonistic because of its allelopathic properties. However, we've had consistent success pairing fennel with head lettuce and some success with celery. Dill is often recommended; dill and fennel are both in the Apiaceae family, which doesn't have to be a negative, but the plants do share common pests and diseases. Chicory is sometimes suggested as a companion.

GARLIC

Critical Period: Garlic is not a strong competitor. The estimated critical period for garlic is 21 to 49 days after emergence.[12] That is not very helpful information for me, though, as garlic emerges in November at Rough Draft Farmstead. Instead, we simply start the emergence count in mid-February when our garlic is beginning to grow rapidly.

Pairings: Though there are reports of green garlic providing some benefits when paired with cucumbers, we have not been pleased with any garlic interplants we have tried. Garlic does not compete well when growing beside another crop even with very little crowding—the plants grow vigorously but the bulbs do not size up well. Instead, consider garlic for relay cropping with cucurbits or planting beneath non-overwintering fall crops like broccoli or cauliflower. Strawberries are a common pairing in a relay.

GINGER

Critical Period: Ginger is a poor competitor. In our experience, though, if given enough space, ginger can handle a small amount of competition and still yield well. Keeping ginger competition-free for 30 days post emergence is advised.[13]

Pairings: We have found that beets or lettuce work well with ginger, as long as spacing is sufficiently generous. Ginger tolerates a small amount of shade, and a light sowing of cosmos, zinnias, or similar annual flowers as a companion crop is a way to maximize return from the bed. Ginger is slow to sprout in the spring, and planting an early run of spring radishes before emergence can add profitability and diversity to ginger beds.

KALE

Critical Period: No clearly defined critical period for kale has been established, but gleaning from information about other brassica crops, it is safest to keep kale competition-free for the first six weeks after emergence and to not pair it with another long-season crop.

Pairings: Green onions can be transplanted (spaced 6 inches [15 cm] or more apart) into widely spaced stands of transplanted kale. As long as the kale rows are at least 18 inches (45 cm) apart, this is a great way to use the space between rows. Parsley, peas, cilantro, and carrots can also be undersown as a way to provide microbial support, but likely not for a full crop. Overseed cilantro or parsley in the fall.

LEEKS

Critical Period: Leeks are a medium competitor. They should remain competition-free for 60 to 80 days past emergence to avoid significant losses in yield.[14]

Pairings: We overwinter leeks and have had success with a fast round of radishes in the spring to take advantage of the slow maturation and short height of the spring radish greens. Singing Frogs Farm told me they have achieved as many as three rounds of romaine through a single leek bed in the spring. Celery, which has a strong lateral rooting system, has been shown to be an effective pairing with one foot (30 cm) of spacing between celery and leeks—a deep rooter.

LETTUCE

Critical Period: Lettuce is a strong competitor. Although it prefers to remain competition-free for four weeks after transplant for best yields, we and many other growers have observed that, as long as adequate spacing is allowed, lettuce is among the best interplanting options.[15]

Pairings: Because of its fast growth and relatively shallow roots, head lettuce is useful for pairing with many crops, including tomatoes, peppers, eggplant, fennel, leeks, celery, radishes, green onions, okra, most annual herbs, and cucumbers.

OKRA

Critical Period: The most important time to avoid competition with okra is six to eight weeks after sowing. We transplant okra around four weeks after emergence (it does not do well if left in cells or soil blocks for very long).[16]

Pairings: Okra works well with beets, head lettuce, cilantro, sweet potatoes, turnips, or radishes. Can be transplanted into a mature stand of carrots as well in a relay scenario. Undersow with rye and vetch in the late summer.

ONIONS, BULB

Critical Period: For the first two weeks after transplant, onions can handle mild competition with minimal yield loss, but for sizable bulbs they require next to no competition as they mature. For that reason, I don't suggest bulb onions in most interplanting scenarios.[17]

Pairings: Onions are often planted as pest deterrents, though it is not common to get a good crop when bulb onions are interplanted into, say, broccoli. I suggest keeping onions well-spaced from other crops in the bed when interplanting—poor air circulation invites disease and reduces bulb size.

ONIONS, GREEN

Critical Period: Green onions are decent competitors as long as a large bulb is not desired. In fact, some degree of competition drives onion plants to put more energy into creating substantial leaves, which can be a desirable trait depending on the market.

Pairings: Green onions get along with most every crop and most annual herbs such as dill and basil. They work very well with most long-season crops including nightshades and slower-growing brassicas such as broccoli and cauliflower. If spacing allows, a transplanted row of green onions between kale can boost bed profitability. Starting green onions in blocks or cells two weeks before lettuce and transplanting them together is becoming standard practice at our farm—the lettuce leaves can even help blanch the immature "bulbs."

PARSNIP

Critical Period: I have not found any definitive research on competition for parsnips, but parsnips are in the Apiaceae family, and those crops generally do not handle competition well. This rings true of parsnips in our experience.

Pairings: Parsnips are an extremely long-season crop (we sow in late winter and harvest in mid-fall, nine months later). Germination and emergence can take a full month. I advise keeping a parsnip crop free of competition for the majority of its life. In our experience, quick brassicas like radishes do not do well with parsnips. In 2020 we allowed common yellow wood sorrel (*Oxalis stricta*) to continue growing below the parsnips to no obvious ill effect. The sorrel eventually became shaded out by the parsnip greens.

PEAS

Critical Period: Peas are not strong competitors.[18] Ideally, minimize competition between 20 and 70 days after sowing or between emergence and six nodes—whichever comes first—although they can handle competition even during that time.[19] For vining peas, their height gives them a physical advantage in an interplanted bed. Bush peas should be given even less competition.

Pairings: Peas, like most legumes, get along with many other crops because they generate their own nitrogen supply through their association with nitrogen-fixing bacteria. Our favorite pairing is spinach—both crops do well in the early spring (we direct sow spinach around the winter solstice and peas in February). Peas also pair reasonably well with corn, green onions, chicory, arugula, and most brassicas.

PEPPERS

Critical Period: Crops in the Solanaceae family are fairly strong competitors. The research suggests that they will perform best if they remain competition-free for 5 to 10 weeks after transplant. That said, yield losses are minimal, at around 5 percent under medium competition.[20]

Pairings: All the short, fast-growing crops including lettuce, green onions, beets, turnips, arugula, and radishes work well with peppers. The only long-season crop we have had good success with as a pepper companion is celery. Celeriac was not as successful when we transplanted it to share a bed with peppers.

POTATOES

Critical Period: Potatoes are fairly strong competitors. They prefer 8 to 60 days competition-free after emergence, but as long as said competition is not too dense or around for too long the losses are low, in the 5 percent range.[21]

Pairings: The usual suspects of lettuce, beets, radishes, turnips, and green onions make excellent potato bed pairings, as can a sparse planting of annual flowers like zinnia, marigolds, cosmos, or flowering dill. Potatoes work well in relay cropping with squash, but not with crops such as corn or okra that have a long growing season.

RADISHES

Critical Period: There is a critical period of competition for radishes between 13 and 21 days after emergence, but radishes are a fast grower and an excellent competitor.[22] There is little to worry about here, except that in dense plantings the greens grow more robust than the roots, which can hurt marketability.

Pairings: There are only a few crops, even within the brassica family, that can't be combined with a quick round of radishes. Green onions, zucchini, lettuce, cucumbers, broccoli, tomatoes, peppers, eggplant, and slow-growing flowers all work well with radishes. Parsnips and garlic are the only two crops I have not had good luck pairing with radishes. Sowing radishes with carrots is sometimes recommended by gardeners, but we have seen a dramatic loss in root size for both crops in the few times we have tried this pairing.

SPINACH

Critical Period: Spinach is a relatively poor competitor when transplanted between crops, but it is a decent companion when given its own rows next to another crop. Otherwise, it should be kept relatively competition-free until mature.

Pairings: Planting half of a bed with spinach and half of a bed with a legumes, lettuce, or allium has been a good strategy for us. Daisy-family flowers can serve well as companions, too. Spinach does not make mycorrhizal associations, so relay cropping spinach with green onions or lettuce can provide some nutritional advantages.

SQUASH, SUMMER

Critical Period: Summer squash are not strong competitors because their roots are sensitive to disturbance. With that in mind, however, summer squash, including zucchini, can work well as relay crops, as long as the preceding crop is harvested within a couple of weeks after planting the squash.[23]

Pairings: Green onions, potatoes, beets, radishes, turnips, and carrots work as relay crops. A thin stand of flowers such as dill, zinnias, cosmos, or cilantro can provide light shade in the summer and call in beneficials.

SQUASH, WINTER

Critical Period: Keep competition-free for up to four weeks after emergence. However, as long as you can provide sufficient space, winter squashes and pumpkins combine well with taller crops such as corn or sunflowers.

Pairings: Corn, tall legumes (edamame work well), and daisy-family flowers are the best options. Undersow with green onions, parsley, or carrots, but note that the role of these companion plants is to help feed the soil. They may not develop into a salable crop.

THE DAISY FAMILY (ASTERACEAE)

Critical Period: The daisy family includes some fairly strong competitors such as zinnias, daisies, sunflowers, and lettuce. Ideally, keep them competition-free for up to nine weeks, though yield losses are minimal as long as the pairing does not shade the flowers too greatly.[24]

Pairings: Zinnias, marigolds, cosmos, and others are excellent interplanting crops for corn, cucumbers, beans, sweet potatoes, nightshades, perennial herbs, borage, and winter and summer squashes. Sunflowers work well with lettuce, arugula, beets, radishes, green onions, and turnips.

SWEET POTATOES

Critical Period: Sweet potatoes prefer two to six weeks without competition after transplant.[25]

Pairings: Okra and dent corn are both good pairings. Peanuts are not a common crop in market gardens, but they do pair well with sweet potatoes as long as both crops are given enough room to grow. Tall legumes like edamame can also work well, as can tall daisy-family plants.

TOMATOES

Critical Period: Although tomatoes are generally strong competitors, most intercrops should be removed or replaced within five or six weeks. Also, once tomato plants bush out, there are few crops that can successfully grow beneath them without the tomatoes being heavily pruned.

Pairings: Beets, bush beans, green onions, lettuce, carrots, chicory, arugula, agretti, turnips, and radishes are all excellent with tomatoes. We have also had decent yields with basil and cilantro. Borage, sweet alyssum, nasturtium, marigolds, and other bushier varieties of flowers are also excellent pairings in tomato beds.

TURNIPS

Critical Period: My review of the research did not reveal any information about a specific critical period of competition for turnips, but in our experience they are not as competitive as other crops, nor as fast growing as radishes.

Pairings: Turnips are nonmycorrhizal, so pairing them with a crop such as green onions or lettuce, which does form mycorrhizal associations, is a good place to start. Solanaceous crops and cucurbits work well. Turnips work distinctly well when sown with fall cover crops. In that space-saving scenario they can be harvested out of the cover crop as a cash crop without taking up bed space elsewhere. In our trials, turnips have not proved to be a good pairing with parsnips.

Resources and Recommended Reading

I believe wholeheartedly in sharing the ideas and work of others, and in that spirit I share these listings of resources where I seek out no-till information.

MAGAZINES AND ONLINE RESOURCES

Growing for Market

This magazine has been putting out no-till information since long before it was cool. Editor Andrew Mefferd is not only a no-till practitioner himself but has also published a book on the subject, *The Organic No-Till Farming Revolution.*

ACRES U.S.A.

Another magazine that has published numerous articles on soil management and no-till. It's an excellent resource all around, and their yearly Eco-Ag Conference is one of the best in the United States.

Farm Small Farm Smart, *Permaculture Voices*, and *In Search of Soil*

These three podcasts run by Diego Footer have introduced me to a number of great soil management and composting practices, and much more.

Regenerative Agriculture Podcast

John Kempf is a wealth of knowledge, and he dives into soil science with various researchers and farmers on this excellent podcast.

No-till Farmer

This is a podcast dedicated to large-scale conventional agriculture, but it has produced many conversations from which I have gleaned useful ideas and information.

Priming for Production

Natalie Lounsbury was the project leader for this excellent podcast series on soil organic matter, detailing how soil organic matter works, and why it's so important.

Farmer to Farmer Podcast

The late, great Chris Blanchard's podcast remains one of the most valuable resources on the web. There are over 170 interviews to listen to; several explore no-till and ecological farming practices in vegetable production.

Instagram and Facebook

For all their flaws, these two sites are rich with information about growing. Facebook has several useful groups including The No-Till Growers Group and the Market Gardening Success Group, Many of the growers mentioned in this book maintain profiles on one or both of these sites, where they often demonstrate their work.

YouTube

There are so many great farmers who regularly post on YouTube, including Josh Sattin, Richard Perkins, and Charles Dowding. No-Till Growers has a YouTube channel with hours of helpful video. There are lectures as well from people like Dr. Christine Jones and Dr. Elaine Ingham on soil health and biology.

Online Courses

Many growers have started courses that are worth exploring. No-Till Growers is currently working on a free online academy that is slated to launch in late 2021. Among courses currently available, I recommend:

- The Market Gardener's Masterclass from Jean-Martin Fortier
- The Lettuce Masterclass from Ray Tyler
- Neversink Online Market Farming Course from Conor Crickmore
- Soil Food Web School from Dr. Elaine Ingham

BOOKS

JADAM Organic Farming **by Youngsang Cho (JADAM, 2016)**
An excellent expansion of Korean Natural Farming (KNF) principles, especially regarding disease and pest management.

The New Organic Grower **by Eliot Coleman (Chelsea Green Publishing, 2018)**
One of the most comprehensive books on small-scale vegetable production. It laid down the basis for the current organic market gardening movement.

Organic Gardening: The Natural No-Dig Way **by Charles Dowding (UIT Cambridge Ltd.; 3rd edition, 2013)**
A superb introduction to Dowding's straightforward no-till (or as he calls it, no-dig) methods with deep compost mulching.

Farming for the Long Haul **by Michael Foley (Chelsea Green Publishing, 2019)**
A big-picture look at the agricultural practices that will and won't sustain into the future.

The One-Straw Revolution **by Masanobu Fukuoka (The New York Review of Books, 2009)**
My first introduction into "do nothing" farming and no-till practices—an absolute classic.

The Market Gardener **by Jean-Martin Fortier (New Society Publishers, 2014)**
A seminal work in which Fortier describes the small-scale, high-profit market gardening system that he and his wife Maude-Hélène Desroches have employed for decades in Quebec.

***Teaming with Microbes*, *Teaming with Nutrients*, and *Teaming with Fungi* by Jeff Lowenfels (Timber Press)**
This trilogy is an essential primer when it comes to understanding the complex interactions among plant roots, microbes, and nutrients.

The No-Till Organic Vegetable Farm **by Daniel Mays (Storey Publishing, 2020)**
Frith Farm has some of the best and most refined no-till systems out there, and Mays does an absolutely excellent detailing of them in this book.

The Organic No-Till Farming Revolution **by Andrew Mefferd (New Society Publishers, 2019)**
One of the first books on small-scale no-till methods; Mefferd profiles 17 different growers and gives numerous useful details about each of their systems.

Growing a Revolution **by David R. Montgomery (W. W. Norton & Company, 2017)**
Montgomery explores fallen societies throughout history and how the agricultural practices that led to those downfalls mimic how much of agriculture today is currently treating the soil.

No-Till Intensive Vegetable Culture **by Bryan O'Hara (Chelsea Green Publishing, 2020)**
O'Hara has been developing the no-till systems at Tobacco Road Farm in Connecticut for many years, and this book masterfully invites you to explore his practices for use on your own farm.

Farming While Black **by Leah Penniman (Chelsea Green Publishing, 2018)**
One of the best resources for understanding the challenges of Black farmers, as well as a superb

guide for many agricultural contributions of Black land stewards throughout history.

A Soil Owner's Manual **by Jon Stika (Self-published, 2016)**
Absolutely one of the best introductions into soil management, and a must-read for anyone hoping to better understand how to take care of their soil.

The Regenerative Growers Guide to Garden Amendments **by Nigel Palmer (Chelsea Green Publishing, 2020)**
An excellent introduction into making your own garden amendments from what you have available using techniques derived largely from Korean Natural Farming (KNF).

Lastly, the best place to learn how to manage living soil is from the soil itself. No book or video or article will ever teach you as much about how to garden as Nature will. So be open and be observant. Share your results, and don't hide your failures. And remember that being a good grower is really not about growing anything at all—it's about bringing your soil to life, and doing what you can to keep it there.

TOOL MANUFACTURERS AND DISTRIBUTORS

BCS America
www.bcsamerica.com

Community Machinery
www.instagram.com/communitymachinery

Deerfield Supplies
2820 Stringtown Road, Elkton, KY 42220
(270) 265-2425

Earth Tools BCS
www.earthtools.com

Farmers Friend
www.farmersfriend.com

Growers & Co.
www.growers.co

Hoss
www.hosstools.com

Johnny's Selected Seeds
www.johnnyseeds.com

Landzie
www.landzie.com

Neversink Farm Tools
www.neversinktools.com

Paperpot Co
www.paperpot.co

Rimol Greenhouse Systems
www.rimolgreenhouses.com

SEED COMPANIES

Ann Arbor Seed Company (Green Things Farm Collective)
www.a2seeds.com

Fedco Seeds
www.fedcoseeds.com

High Mowing Organic Seeds
highmowingseeds.com

Johnny's Selected Seeds
www.johnnyseeds.com

Kitazawa Seed Co.
www.kitazawaseed.com

Osborne Quality Seeds
www.osborneseed.com

San Diego Seed Company
www.sandiegoseedcompany.com

Southern Exposure Seed Exchange
www.southernexposure.com

Territorial Seed Company
www.territorialseed.com

The Buffalo Seed Company
www.thebuffaloseedcompany.com
Tourne-Sol Co-operative Farmboutique.
fermetournesol.qc.ca

Wild Mountain Seeds
www.wildmountainseeds.com

Notes

Chapter One: The Basic Science of Living Soil

1. Todd B. Bates, "How Plants Harness Microbes to Get Nutrients," Phys.org, September 17, 2018, https://phys.org/news/2018-09-harness-microbes-nutrients.html.
2. Torgny Näsholm et al., "Uptake of Organic Nitrogen by Plants," *New Phytologist* 182 (2009): 31–48, https://doi.org/10.1111/j.1469-8137.2008.02751.x.
3. Thimmaraju Rudrappa et al., "Root-Secreted Malic Acid Recruits Beneficial Soil Bacteria," *Plant Physiology* 148 (November 2008): 1547–56, http://doi.org/10.1104/pp.108.127613.
4. Alan R. Putnam, Joseph Defrank, and Jane P. Barnes, "Exploitation of Allelopathy for Weed Control in Annual and Perennial Cropping Systems," *Journal of Chemical Ecology* 9 (August 1983): 1001–10, https://doi.org/10.1007/BF00982207.
5. Hina Upadhyay, Vipul Kumar, and Sanjeev Kumar, "The Effect of Fungal Metabolites on Soil-Borne Pathogenic Fungi," in *New and Future Developments in Microbial Biotechnology and Bioengineering*, edited by Joginder Singh and Praveen Gehlot (Amsterdam: Elsevier, 2020).
6. David M. Weller et al., "Microbial Populations Responsible for Specific Soil Suppressiveness to Plant Pathogens," *Annual Review of Phytopathology* 40, no. 1 (2002): 309–48, https://doi.org/10.1146/annurev.phyto.40.030402.110010.
7. Paula Muñoz and Sergi Munné-Bosch, "Photo-Oxidative Stress during Leaf, Flower and Fruit Development," *Plant Physiology* 176 (February 2018): 1004–14, https://doi.org/10.1104/pp.17.01127.
8. Joseph Masabni et al., "Shade Effect on Growth and Productivity of Tomato and Chili Pepper," *HortTechnology* 26, no. 3 (June 2016): 344–50, https://doi.org/10.21273/HORTTECH.26.3.344.
9. Hongfang Zhu et al., "Effects of Low Light on Photosynthetic Properties, Antioxidant Enzyme Activity, and Anthocyanin Accumulation in Purple Pak-Choi (Brassica campestris ssp. Chinensis Makino)," PLoS ONE 12, no. 6 (2017): e0179305, https://doi.org/10.1371/journal.pone.0179305.
10. Leiv M. Mortensen, "Review: CO_2 Enrichment in Greenhouses. Crop Responses," *Scientia Horticulturae* 33, nos. 1–2 (August 1987): 1–25, https://doi.org/10.1016/0304-4238(87)90028-8.
11. Megha Poudel and Bruce Dunn, "Greenhouse Carbon Dioxide Supplementation," Oklahoma State University, March 2017, https://extension.okstate.edu/fact-sheets/greenhouse-carbon-dioxide-supplementation.html.
12. Hadi Pirasteh-Anosheh et al., "Stomatal Responses to Drought Stress," in *Water Stress and Crop Plants: A Sustainable Approach*, ed. Parvaiz Ahmad (Hoboken, NJ: John Wiley & Sons, 2016).
13. T. T. Kozlowski, "Plant Responses to Flooding of Soil," *BioScience* 34, no. 3 (March 1984): 162–67, https://doi.org/10.2307/1309751.

14. Kyaw Aung, Yanjuan Jiang, and Sheng Yang He, "The Role of Water in Plant–Microbe Interactions," *Plant Journal* 93, no. 4 (February 2018): 771–80, https://doi.org/10.1111/tpj.13795.
15. Randy Ortíz-Castro et al., "The Role of Microbial Signals in Plant Growth and Development," *Plant Signaling and Behavior* 4, no. 8 (August 2009): 701–12, https://doi.org/10.4161/psb.4.8.9047.
16. Robin Heinen et al., "Effects of Soil Organisms on Aboveground Plant-Insect Interactions in the Field: Patterns, Mechanisms and the Role of Methodology," *Frontiers in Ecology and Evolution* 24, (July 2018): article 106, https://doi.org/10.3389/fevo.2018.00106.
17. Nicholas A. Barber and Nicole L. Soper Gorden, "How Do Belowground Organisms Influence Plant-Pollinator Interactions?" *Journal of Plant Ecology* 8, no. 1 (February 2015): 1–11, https://doi.org/10.1093/jpe/rtu012.
18. Merle Tränkner, Ershad Tavakol, and Bálint Jákli, "Functioning of Potassium and Magnesium in Photosynthesis, Photosynthate Translocation and Photoprotection," *Physiologia Plantarum* 163, no. 3 (2018): 414–31, https://doi.org/10.1111/ppl.12747.
19. Taka-aki Ono et al., "X-ray Detection of the Period-Four Cycling of the Manganese Cluster in Photosynthetic Water Oxidizing Enzyme," *Science* 258, no. 5086 (November 1992): 1335–37, https://doi.org/10.1126/science.258.5086.1335.
20. Benjamin I. Cook, Richard Seager, and Jason E. Smerdon, "The Worst North American Drought Year of the Last Millennium: 1934," *Geophysical Research Letters* 41, no. 20 (2014): 7298–305, https://doi.org/10.1002/2014GL061661.
21. Timothy Egan, *The Worst Hard Time: The Untold Story of Those Who Survived the Great American Dust Bowl* (Boston: Mariner Books, 2006).
22. Andrew Lambert et al., "Dust Impacts of Rapid Agricultural Expansion on the Great Plains," *Geophysical Research Letters* 47, no. 20 (2020): e2020GL090347, https://doi.org/10.1029/2020GL090347.
23. D. L. Evans et al., "Soil Lifespans and How They Can Be Extended by Land Use and Management Change," *Environmental Research Letters* 15, no. 9 (September 2020), https://doi.org/10.1088/1748-9326/aba2fd.
24. Chris Arsenault, "Only 60 Years of Farming Left If Soil Degradation Continues," *Scientific American*, December 5, 2014, https://www.scientificamerican.com/article/only-60-years-of-farming-left-if-soil-degradation-continues.
25. Dixie Sandborn, "Weeds Are an Indicator of a Soil's Health," Michigan State University Extension, August 15, 2016, https://www.canr.msu.edu/news/weeds_are_an_indicator_of_a_soils_health.
26. Sjoerd Willem Duiker, "Effects of Soil Compaction," Penn State University, March 8, 2005, https://extension.psu.edu/effects-of-soil-compaction.
27. Søren Navntoft et al., "Weed Seed Predation in Organic and Conventional Fields," *Biological Control* 49, no. 1 (2009): 11–16, https://doi.org/10.1016/j.biocontrol.2008.12.003.
28. Liz Mineo, "Good Genes are Nice, But Joy Is Better," *Harvard Gazette*, April 11, 2017, https://news.harvard.edu/gazette/story/2017/04/over-nearly-80-years-harvard-study-has-been-showing-how-to-live-a-healthy-and-happy-life.

Chapter Three: Compost in the No-Till Garden

1. M. B. Whitt and C. S. Coker, "Implementing a Plant Growth Testing Program," US Composting Council, 2015, https://cdn.ymaws.com/www.compostingcouncil.org/resource/resmgr/images/USCC-PH-Fact-Sheet-3-for-web.pdf.

CHAPTER FOUR: MULCH

1. "The Peatlands Stewardship and Related Amendments Act," The Legislative Assembly of Manitoba, 3rd Session, 40th Legislature, https://web2.gov.mb.ca/bills/40-3/b061e.php.
2. Suresh K. Malhotra, "Diagnosis and Management of Soil Fertility Constraints in Coconut (*Cocos nucifera*): A Review," Indian Journal of Agricultural Sciences 87, no. 6 (June 2017): 711–26, https://www.researchgate.net/publication/317568008_Diagnosis_and_management_of_soil_fertility_constraints_in_coconut_Cocos_nucifera_A_review.
3. Dunmei Lin et al., "Microplastics Negatively Affect Soil Fauna but Stimulate Microbial Activity: Insights from a Field-Based Microplastic Addition Experiment," *Proceedings of the Royal Society B* 287, no. 1934 (September 2020), https://doi.org/10.1098/rspb.2020.1268.

CHAPTER FIVE: TURNING OVER BEDS

1. Amy Westervelt, "Phthalates Are Everywhere, and the Health Risks Are Worrying. How Bad Are They Really?" *Guardian*, February 10, 2015, https://www.theguardian.com/lifeandstyle/2015/feb/10/phthalates-plastics-chemicals-research-analysis.

CHAPTER SEVEN: FERTILITY MANAGEMENT

1. Stefano Manzoni, Joshua P. Schimel, and Amilcare Porporato, "Responses of Soil Microbial Communities to Water Stress: Results from a Meta-Analysis," *Ecology* 93, no. 4 (April 2012): 930–38, https://doi.org/10.1890/11-0026.1.
2. C. C. G. St. Martin et al., "Effects and Relationships of Compost Type, Aeration and Brewing Time on Compost Tea Properties, Efficacy Against *Pythium ultimum*, Phytotoxicity and Potential as a Nutrient Amendment for Seedling Production," *Biological Agriculture & Horticulture* 28, no. 3 (2012): 185–205, https://doi.org/10.1080/01448765.2012.727667.
3. Christine Jones, "Light Farming: Restoring Carbon, Organic Nitrogen and Biodiversity to Agricultural Soils," AmazingCarbon.com, 2018, https://amazingcarbon.com/JONES-LightFarmingFINAL(2018).pdf.
4. Barry Thornton and David Robinson, "Uptake and Assimilation of Nitrogen from Solutions Containing Multiple N Sources," *Plant, Cell & Environment* 28, no. 6 (March 2005): 813–21.
5. Cai Zejiang et al., "Intensified Soil Acidification from Chemical N Fertilization and Prevention by Manure in an 18-Year Field Experiment in the Red Soil of Southern China," *Journal of Soils and Sediments* 15, (2015): 260–70, https://doi.org/10.1007/s11368-014-0989-y.

CHAPTER EIGHT: TRANSPLANTING AND INTERPLANTING

1. V. Vančura and A. Hovadík, "Root Exudates of Plants: II. Composition of Root Exudates of Some Vegetables," *Plant and Soil* 22, no. 1 (1965): 21–32, https://www.jstor.org/stable/42932087.
2. Erik H. Poelman et al., "Parasitoid-Specific Induction of Plant Responses to Parasitized Herbivores Affects Colonization by Subsequent Herbivores," *PNAS* 108, no. 49 (December 6, 2011): 19647–652, https://doi.org/10.1073/pnas.1110748108.
3. Niall J. A. Conboy et al., "Companion Planting with French Marigolds Protects Tomato Plants from Glasshouse Whiteflies Through the Emission of Airborne Limonene," *PLoS ONE* 14, no. 3 (2019), https://doi.org/10.1371/journal.pone.0213071.
4. Fraide Sulvai, Beni Jequicene Mussengue Chaúque, and Domingos Lusitâneo Pier Macuvele, "Intercropping of Lettuce and

Onion Controls Caterpillar Thread, *Agrotis Ipsilon* Major Insect Pest of Lettuce," *Chemical and Biological Technologies in Agriculture* 3, (2016): article 28, https://doi.org/10.1186/s40538-016-0079-z.

Chapter Nine: Seven No-Till Crops from Start to Finish

1. Travis Cranmer, "The Benefits of Removing the Garlic Scape," Ontario Ministry of Agriculture, Food, and Rural Affairs, updated June 21, 2017, http://www.omafra.gov.on.ca/english/crops/hort/news/hortmatt/2017/10hrt17a1.htm#:~:text=Often%20leaves%20are%20cut%20in,was%20reduced%20by%20approximately%2025%25.

Appendix B: Critical Period of Competition and Interplant Pairings

1. Godfrey G. Maina, Joel S. Brown, and Mordechai Gersani, "Intra-Plant Versus Inter-Plant Root Competition in Beans: Avoidance, Resource Matching or Tragedy of the Commons," *Plant Ecology* 160 (2002): 235–47, https://doi.org/10.1023/A:1015822003011.
2. José Mansilla Martinez et al., "Competition and Critical Periods in Spring Sugar Beet Cultivation," *Journal of Plant Protection Research* 55, no. 4 (2015), https://doi.org/10.1515/jppr-2015-0044.
3. H. Carranza et al., "Determination of Weed Critical Period Competition in Broccoli (*Brassica oleracea* var. italica)," *Integrated Pest Management Collaborative Research Support Program: Fourth Annual Report*, no. 4 (September 1997): 84–88.
4. Clarence Swanton et al., "Weed Management in Carrots," Ontario Ministry of Agriculture, Food, and Rural Affairs, 2009, http://www.omafra.gov.on.ca/english/crops/facts/09-045w.htm.
5. J. R. Qasem, "Weed Competition in Cauliflower (Brassica oleracea L. var. botrytis) in the Jordan Valley," *Scientia Horticulturae* 121, no. 3 (2009): 255–59, https://doi.org/10.1016/j.scienta.2009.02.010.
6. D. T. Baumann, "Competitive Suppression of Weeds in a Leek-Celery Intercropping System: An Exploration of Functional Biodiversity," PhD Diss., Wageningen University, January 2001, https://library.wur.nl/WebQuery/wurpubs/108893.
7. Francisco Bedmar, Pablo Manetti, and Gloria Monterubbianesi, "Determination of the Critical Period of Weed Control in Corn Using a Thermal Basis," *Pesquisa Agropecuária Brasileira* 34, no. 2 (1999), http://dx.doi.org/10.1590/S0100-204X1999000200006.
8. Susan E. Weaver, "Critical Period of Weed Competition in Three Vegetable Crops in Relation to Management Practices," *Weed Research* 24, no. 5 (October 1984): 317–25, https://doi.org/10.1111/j.1365-3180.1984.tb00593.x.
9. Xuemei Xiao et al., "Intercropping of Green Garlic (*Allium sativum* L.) Induces Nutrient Concentration Changes in the Soil and Plants in Continuously Cropped Cucumber (*Cucumis sativus* L.) in a Plastic Tunnel," *PLoS ONE* 8, no. 4 (2013): e62173, https://doi.org/10.1371/journal.pone.0062173; Xuemei Xiao et al., "A Green Garlic (*Allium sativum* L.) Based Intercropping System Reduces the Strain of Continuous Monocropping in Cucumber (*Cucumis sativus* L.) by Adjusting the Micro-Ecological Environment of Soil," *PeerJ*, no. 7 (July 2019): e7267, https://doi.org/10.7717/peerj.7267.
10. Alfonso S. Frenda et al., "The Critical Period of Weed Control in Faba Bean and Chickpea in Mediterranean Areas," *Weed Science* 61, no. 3 (2013): 452–59, https://doi.org/10.1614/WS-D-12-00137.1.

11. Khuram Mubeen et al., “Critical Period of Weed-Crop Competition in Fennel (*Foeniculum vulgare* Mill.).” *Pakistan Journal of Weed Science Research* 15, nos. 2–3 (2009): 171–81, https://www.researchgate.net/publication/260294721_CRITICAL_PERIOD_OF_WEED-CROP_COMPETITION_IN_FENNEL_Foeniculum_vulgare_Mill.
12. H. Z. Ghosheh, “Garlic (*Allium sativum*) Response to Weed Control Practices,” *Die Bodenkultur* 51, no. 3 (2000): 157–61, http://citeseerx.ist.psu.edu/viewdoc/download?doi=10.1.1.542.2744&rep=rep1&type=pdf.
13. Habetewold Kifelew, Tadesse Eshetu, and Hailemariam Abera, “Critical Time of Weed Competition and Evaluation of Weed Management Techniques on Ginger (*Zengeber Officinale*) at Tepi in South West Ethiopia,” *International Journal of Research Studies in Agricultural Sciences* 1, no. 3 (2015): 5–10, https://www.researchgate.net/publication/303494824_Critical_Time_of_Weed_Competition_and_Evaluation_of_Weed_Management_Techniques_on_Ginger_Zengeber_Officinale_at_Tepi_in_South_West_Ethiopia.
14. Nihat Tursun et al., “Critical Period for Weed Control in Leek (*Allium porrum* L.),” *HortScience* 42, no. 1 (2007): 106–109, https://doi.org/10.21273/HORTSCI.42.1.106.
15. Sarah R. Parry, Ryan Cox, and Anil Shrestha, “Duration of Weed-Free Period in Organic Lettuce: Crop Yield, Economics, and Crop Quality,” California Weed Science Society Conference Proceedings, 2015, https://www.cwss.org/uploaded/media_pdf/6655-4B_Parry_CWSS%202015_Duration%20of%20Weed-Free%20Period%20in%20Organic%20Lettuce.pdf.
16. Suhair Mohamed Elamin et al., “Critical Period for Weed Control in Okra (*Abelmeschus esculentus* L. *Moench*) in Dongola, Northern State, Sudan,” *Nile Journal for Agricultural Sciences* 4, no. 1 (2019): 15–23.
17. J. R. Qasem, “Critical Period of Weed Competition in Onion (*Allium cepa* L.) in Jordan,” *Jordan Journal of Agricultural Sciences* 1, no. 1 (2010): 32–42, https://www.researchgate.net/publication/273446428_71_Qasem_JR_2005_Critical_period_of_weed_competition_in_onion_Allium_cepa_L_In_Jordan_Jordan_Journal_of_Agricultural_Sciences_JJAS_1_1_32-42.
18. Pamela L. S. Pavek, “Pea (*Pisum sativum* L.),” USDA Natural Resources Conservation Service, updated September 19, 2012, https://www.nrcs.usda.gov/Internet/FSE_PLANTMATERIALS/publications/wapmcfs11388.pdf.
19. K. N. Harker, Robert E. Blackshaw, and George W. Clayton, “Timing Weed Removal in Field Pea (*Pisum Sativum*),” *Weed Technology* 15, no. 2 (2001): 277–83, https://jstor.org/stable/3988804.
20. L. H. O. Uljol et al., “Weed Interference on Productivity of Bell Pepper Crops,” *Planta Daninha* 36, (July 2018), http://dx.doi.org/10.1590/s0100-83582018360100046.
21. Dogan Işik et al., “The Critical Period for Weed Control (CPWC) in Potato (*Solanum tuberosum* L.),” *Notulae Botanicae Horti Agrobotanici Cluj-Napoca* 43, no. 2 (2015): 355–60, https://doi.org/10.15835/nbha43210031.
22. K. D. Harris et al., “Critical Period of Weed Control in Radish (*Raphinus sativus* L.),” *Journal of Agricultural Sciences* 10 (January 2017): 6–10, http://doi.org/10.4038/agrieast.v10i0.23.
23. Lynn M. Sosnoskie, Amy L. Davis, and A. Stanley Culpepper, “Response of Seeded and Transplanted Summer Squash to S-Metolachlor Applied at Planting and Postemergence,” *Weed Technology* 22, no. 2 (2008): 253–56, https://doi.org/10.1614/WT-07-137.1.

24. Gholipour Hossein, et al., "Critical Period of Weeds Control in Sunflower, Helianthus Annus L.," *Agroecology Journal* 5, no. 17 (2010): 75–82, https://www.sid.ir/en/Journal/ViewPaper.aspx?ID=182584.

25. Jessica E. Seem, Nancy G. Creamer, and David W. Monks, "Critical Weed-Free Period for 'Beauregard' Sweetpotato (*Ipomoea batatas*)," *Weed Technology* 17, no. 4 (2003): 686–95, https://doi.org/10.1614/WT02-089.

Bibliography

Acevedo, E., P. Silva, and H. Silva. "Wheat Growth and Physiology." In *Bread Wheat: Improvement and Production*, edited by B. C. Curtis, S. Rajaram, and H. Gómez Macpherson. Rome: Food and Agriculture Organization of the United Nations, 2002. http://www.fao.org/3/y4011e06.htm#bm06.

Albrecht, Julie A. "Escherichia coli O157:H7 (E Coli)." University of Nebraska-Lincoln. Accessed December 22, 2020. https://food.unl.edu/escherichinia-coli-o157h7-e-coli.

Alwell, Audrey. "Sunn Hemp Gains Popularity as a Stress-Tolerant Cover Crop." *Organic Broadcaster*, May/June 2015. https://mosesorganic.org/sunn-hemp/.

Anderson, M. Kat. *Tending the Wild: Native American Knowledge and the Management of California's Natural Resources.* Los Angeles: University of California Press, 2005.

ATTRA. "Can I Use Cardboard and Newspaper as Mulch on My Organic Farm?" National Center for Appropriate Technology. Accessed December 22, 2020. https://attra.ncat.org/can-i-use-cardboard-and-newspaper-as-mulch-on-my-organic-farm/.

Biello, David. "Peat and Repeat: Can Major Carbon Sinks Be Restored by Rewetting the World's Drained Bogs?" *Scientific American*, December 8, 2009. https://www.scientificamerican.com/article/peat-and-repeat-rewetting-carbon-sinks/.

Björkman, T., and J. W. Shail. "Cover Crop Fact Sheet Series, Forage Radish." Cornell University, 2010, Ver. 1.100716. http://www.hort.cornell.edu/bjorkman/lab/covercrops/pdf/radish.pdf.

Bongard, P., E. Oelke, and S. Simmons. "Spring Barley Growth and Development Guide." University of Minnesota Extension. Accessed January 2, 2021. https://extension.umn.edu/growing-small-grains/spring-barley-growth-and-development-guide.

Brainard, Daniel C., R. Edward Peachey, Erin R. Haramoto, John M. Luna, and Anusuya Rangarajan. "Weed Ecology and Nonchemical Management under Strip-Tillage: Implications for Northern U.S. Vegetable Cropping Systems." *Weed Technology* 27, no. 1 (2013): 218–30. https://doi.org/10.1614/WT-D-12-00068.1.

Brown, Gabe. *Dirt to Soil: One Family's Journey into Regenerative Agriculture.* White River Junction, VT: Chelsea Green, 2018.

Caamal-Maldonado, Jesús Arturo, Juan José Jiménez-Osornio, Andrea Torres-Barragán, and Ana Luisa Anaya. "The Use of Allelopathic Legume Cover and Mulch Species for Weed Control in Cropping Systems." *Agronomy Journal* 93, no. 1 (2001): 27–36. https://doi.org/10.2134/agronj2001.93127x.

Caron, Céline. "Ramial Chipped Wood More Than Wood Chips." Maine Organic Farmers and Gardeners Association. Accessed January 2, 2021. https://www.mofga.org/resources/mulch/ramial-chipped-wood-more-than-wood-chips/.

Chalker-Scott, Linda. "The Myth of Pretty Mulch." Washington State University Extension. Accessed January 2, 2021. https://s3.wp.wsu.edu/uploads/sites/403/2015/03/bark-mulch.pdf.

Cho, Youngsang. *JADAM Organic Farming & Gardening Part 1 of 3*. Self-published, 2016.

Clark, Andy, ed. *Managing Cover Crops Profitably, Third Edition*. College Park, MD: Sustainable Agriculture Research and Education, 2007. https://www.sare.org/wp-content/uploads/Managing-Cover-Crops-Profitably.pdf.

Coleman, Eliot. *The New Organic Grower: A Master's Manual of Tools and Techniques for the Home and Market Gardener*. White River Junction, VT: Chelsea Green, 1989.

Curell, Christina. "Champion of Cover Crops: Oats." Michigan State University Extension. September 7, 2012. https://www.canr.msu.edu/news/champion_of_cover_crops_oats.

———. "Mustard as a Cover Crop." Michigan State University Extension. August 22, 2011. https://www.canr.msu.edu/news/mustard_as_a_cover_crop.

Davis, J. G., and P. Kendall. "Preventing *E. coli* from Garden to Plate." Colorado State University Extension, fact sheet 9.369, updated July 2012. https://extension.colostate.edu/docs/foodnut/09369.pdf.

Dawling, Pam. "How and When to Plant Garlic." *Growing for Market*, accessed January 2, 2021. https://www.growingformarket.com/articles/how-and-when-to-plant-garlic.

Dong, Allen. "Dry Farming: Vegetable and Field Crop Rooting Depths and Lateral Distances." Appropriate Technology for Small and Subsistence Farms. Compiled June 2016, revised January 2021. http://members.efn.org/~itech/pdf%20files/Vegetable%20and%20field%20crop%20rooting%20depths%20and%20lateral%20distances.pdf.

Dowding, Charles, and Stephanie Hafferty. *No Dig Organic Home and Garden: Grow, Cook, Use and Store Your Harvest*. East Meon, UK: Permanent Publications, 2017.

Du, Lantian, Baojian Huang, Nanshan Du, Shirong Guo, Sheng Shu, and Jin Sun. "Effects of Garlic/Cucumber Relay Intercropping on Soil Enzyme Activities and the Microbial Environment in Continuous Cropping." *HortScience* 52, no. 1 (2017): 78–84. https://doi.org/10.21273/HORTSCI11442-16.

Duiker, Sjoerd Willem. "Diagnosing Soil Compaction Using a Penetrometer (Soil Compaction Tester)." PennState Extension. Updated January 1, 2002. https://extension.psu.edu/diagnosing-soil-compaction-using-a-penetrometer-soil-compaction-tester.

———. "Management of Red Clover as a Cover Crop." PennState Extension. Updated October 30, 2007. https://extension.psu.edu/management-of-red-clover-as-a-cover-crop.

Fidanza, Mike, and David Beyer. "Plant Nutrients and Fresh Mushroom Compost." Full Circle Mushroom Compost, LLC. Accessed January 2, 2021. http://www.fullcirclemushroomcompost.com/wp-content/uploads/2016/12/Fidanza-BeyerNutrient.pdf.

Folds, Evan. "The Ultimate Guide to Brewing Compost Tea." Medium, March 10, 2018. https://medium.com/@evanfolds/the-ultimate-guide-to-brewing-compost-tea-8ddeb4622b9e.

Foley, Michael. *Farming for the Long Haul: Resilience and the Lost Art of Agricultural Inventiveness*. White River Junction, VT: Chelsea Green, 2019.

Food and Agriculture Organization of the United Nations. "Conservation Agriculture." Accessed January 2, 2021. www.fao.org/3/a-i7480e.pdf.

———. "Physical Factors Affecting Soil Organisms." Accessed January 2, 2021. http://www.fao.org/agriculture/crops/thematic-sitemap/theme/spi/soil-biodiversity/soil-organisms/physical-factors-affecting-soil-organisms/en/.

Fortier, Jean-Martin. *The Market Gardener: A Successful Grower's Handbook for*

Small-Scale Organic Farming. Gabriola Island, BC: New Society, 2014.

Fukuoka, Masanobu. *The One-Straw Revolution: An Introduction to Natural Farming*. The New York Review of Books, Inc., 2009.

Gaur, Nisha, Aayush Kukreja, Mahavir Yadav, and Archana Tiwari. "Assessment of Phytoremediation Ability of *Coriander sativum* for Soil and Water Co-Contaminated with Lead and Arsenic: A Small-Scale Study." *3 Biotech* 7, no. 3 (2017): article 196. https://doi.org/10.1007%2Fs13205-017-0794-6.

Gliessman, Stephen R., and Miguel A. Altieri. "Polyculture Cropping has Advantages." *California Agriculture* 36, no. 7 (1982): 14–16. http://calag.ucanr.edu/archive/?type=pdf&article=ca.v036n07p14.

Griffiths, Georgianne J. K., Linton H. Winder, John M. Holland, C. F. George Thomas, and Eirene Williams. "The Representation and Functional Composition of Carabid and Staphylinid Beetles in Different Field Boundary Types at a Farm-Scale." *Biological Conservation* 135, no. 1 (2007): 145–52. https://doi.org/10.1016/j.biocon.2006.09.016.

Hauggaard-Nielsen, H., and E. S. Jensen. "Facilitative Root Interactions in Intercrops." *Plant and Soil* 274, (2005): 237–50. https://doi.org/10.1007/s11104-004-1305-1.

Hollister, Julia. "Sunflowers Could Be an Easy, Water-Saving Cover Crop." *Farm Progress*, June 19, 2018. https://www.farmprogress.com/tree-nuts/sunflowers-could-be-easy-water-saving-cover-crop.

Hurt, R. Douglas. *Indian Agriculture in America: Prehistory to the Present*. Lawrence: University Press of Kansas, 1988.

Hyland, Charles, Quirine Ketterings, Dale Dewing, Kristin Stockin, Karl Czymmek, Greg Albrecht, and Larry Geohring. "Phosphorus Basics—The Phosphorus Cycle." Cornell University Cooperative Extension. Accessed January 2, 2021. http://nmsp.cals.cornell.edu/publications/factsheets/factsheet12.pdf.

Isleib, Jim. "Soil Temperature, Seed Germination and the Unusual Spring of 2012." Michigan State University Extension. May 4, 2012. https://www.canr.msu.edu/news/soil_temperature_seed_germination_and_the_unusual_spring_of_2012.

Itani, T., Y. Nakahata, and H. Kato-Noguchi. "Allelopathic Activity of Some Herb Plant Species." *International Journal of Agriculture and Biology* 15, no. 6. (2013): 1359–62. https://www.researchgate.net/publication/287756602_Allelopathic_activity_of_some_herb_plant_species.

Iverson, Aaron L., Linda E. Marín, Katherine K. Ennis, David J. Gonthier, Benjamin T. Connor-Barrie, Jane L. Remfert, Bradley J. Cardinale, and Ivette Perfecto. "Review: Do Polycultures Promote Win–Wins or Trade–Offs in Agricultural Ecosystem Services? A Meta–Analysis." *Journal of Applied Ecology* 51, no. 6 (2014): 1593–602. https://doi.org/10.1111/1365-2664.12334.

Jacobs, Alayna. "Plant Guide for Oilseed Radish (*Raphanus sativus* L.)." Booneville, AR: USDA-Natural Resources Conservation Service, 2012. https://www.nrcs.usda.gov/Internet/FSE_PLANTMATERIALS/publications/arpmcpg11828.pdf.

Johanowicz, Denise L., and Everett R. Mitchell. "Effects of Sweet Alyssum Flowers on the Longevity of the Parasitoid Wasps Cotesia Marginiventris (Hymenoptera: Braconidae) and Diadegma Insulare (Hymenoptera: Ichneumonidae)." *Florida Entomologist* 83, no. 1 (2000): 41–47. https://journals.flvc.org/flaent/article/view/59514.

Kandel, Hans. "Flax Production (4/26/12)." Crop and Pest Report. North Dakota State University, April 26, 2012. https://www.ag.ndsu.edu/cpr/plant-science/flax-production-4-26-12.

Kilian, Robert. "Lacy Phacelia: *Phacelia Tanacetifolia* Benth: A Native Annual Forb for Conservation Use in Montana and

Wyoming." Natural Resources Conservation Service, Plant Materials Technical Note, August 2016. https://www.nrcs.usda.gov/Internet/FSE_PLANTMATERIALS/publications/mtpmctn12938.pdf.

Kulikoswksi, Mick. "Sunflower Pollen Has Medicinal, Protective Effects on Bees." North Carolina State University, September 26, 2018. https://news.ncsu.edu/2018/09/sunflower-pollen-protects-bees/.

Lehmann, J., and G. Schroth. "Nutrient Leaching." In *Trees, Crops, and Soil Fertility: Concepts and Research Methods*, edited by Fergus L. Sinclair and Goetz Schroth. Wallingford, UK: CABI Publishing, 2003. 151–66. http://www.css.cornell.edu/faculty/lehmann/publ/Lehmann%20et%20al.,%202003,%20Leaching%20CABI%20book.pdf.

Leskovar, Daniel I., Daniel J. Cantliffe, and Peter J. Stoffella. "Growth and Yield of Tomato Plants in Response to Age of Transplants." *Journal of the American Society for Horticultural Science* 116, no. 3 (1991): 416–20. https://doi.org/10.21273/JASHS.116.3.416.

Lindemann, W. C. and C.R. Glover. Revised by Robert Flynn and John Idowu. "Nitrogen Fixation by Legumes." College of Agricultural, Consumer and Environmental Sciences, New Mexico State University. Updated June 2015. https://aces.nmsu.edu/pubs/_a/A129/.

Lowenfels, Jeff. *Teaming with Nutrients: The Organic Gardener's Guide to Optimizing Plant Nutrition*. Portland, OR: Timber Press, 2013.

———. *Teaming with Fungi: The Organic Grower's Guide to Mycorrhizae*. Portland, OR: Timber Press, 2017.

Lowenfels, Jeff, and Wayne Lewis. *Teaming with Microbes: The Organic Gardener's Guide to the Soil Food Web*. Revised edition. Portland, OR: Timber Press, 2010.

Magdoff, Fred, and Harold van Es. *Building Soils for Better Crops: Sustainable Soil Management*. Third Edition. Brentwood, MD: SARE Outreach Publications, 2009. https://www.sare.org/wp-content/uploads/Building-Soils-For-Better-Crops.pdf.

Mays, Daniel. *The No-Till Organic Vegetable Farm: How to Start and Run a Profitable Market Garden That Builds Health in Soil, Crops, and Communities*. North Adams, MA: Storey Publishing, 2020.

McGuire, Andy. "Sudangrass and Sorghum-Sudangrass Hybrids." Washington State University Extension, 2003. https://www.canr.msu.edu/uploads/234/78912/EB1950E.pdf.

Mefferd, Andrew. *The Organic No-Till Farming Revolution: High-Production Methods for Small-Scale Farmers*. Gabriola Island, BC: New Society, 2019.

———. *The Greenhouse and Hoophouse Growers Handbook: Organic Vegetable Production Using Protected Culture*. White River Junction, VT: Chelsea Green, 2017.

Mohler, Charles L., and Sue Ellen Johnson. *Crop Rotation on Organic Farms: A Planning Manual*. Brentwood, MD: SARE Outreach Publications, 2020. https://www.sare.org/wp-content/uploads/Crop-Rotation-on-Organic-Farms.pdf.

Montgomery, David R. *Growing a Revolution: Bringing Our Soil Back to Life*. New York: W. W. Norton & Company, 2017.

Morandin, Lora, Rachael F. Long, Corin Pease, and Claire Kremen. "Hedgerows Enhance Beneficial Insects on Farms in California's Central Valley." *California Agriculture* 65, no. 4 (2011): 197–201. https://doi.org/10.3733/ca.v065n04p197.

Nair, P. K. R., B. T. Kang, and D. C. L. Kass. "Nutrient Cycling and Soil–Erosion Control in Agroforestry Systems." In *Agriculture and the Environment: Bridging Food Production and Environmental Protection in Developing Countries, Volume 60*, edited by Anthony S. R. Juo and Russell D. Freed, 117–38. Madison, WI: American Society of Agronomy, Crop Science Society of America, and Soil Science Society of America, 1995). https://doi.org/10.2134/asaspecpub60.c7.

Narla, R. D, J. W. Muthomi, S. M. Gachu, J. H. Nderitu, and F. M. Olubayo. "Effect of Intercropping Bulb Onion and Vegetables on Purple Blotch and Downy Mildew." *Journal of Biological Sciences* 11, no. 1 (2011): 52–57. https://doi.org/10.3923/jbs.2011.52.57.

Newton, Blake. "Assassin Bugs and Ambush Bugs." Kentucky Critter Files. University of Kentucky Department of Entomology. Updated October 19, 2006. https://www.uky.edu/Ag/CritterFiles/casefile/insects/bugs/assassin/assassin.htm.

Nishida, Florence. "Vegetable Root Depth to Gauge Watering Depth." University of California Cooperative Extension, September 2011. http://celosangeles.ucanr.edu/files/121762.pdf.

Nordlund, Donald A., Richard B. Chalfant, and W. J. Lewis. "Arthropod Populations, Yield and Damage in Monocultures and Polycultures of Corn, Beans and Tomatoes." *Agriculture, Ecosystems & Environment* 11, no. 4 (1984): 353–67. https://doi.org/10.1016/0167-8809(84)90007-0.

Norwegian University of Science and Technology. "Peat Moss, a Necessary Bane." Phys.org. May 21, 2015. https://phys.org/news/2015-05-peat-moss-bane.html.

NRCS. "Plant Fact Sheet: Hairy Vetch *Vicia villosa* Roth." National Resource Conservation Service. Updated February 5, 2002. https://plants.usda.gov/factsheet/pdf/fs_vivi.pdf.

O'Briant, Melinda, and Katie Charlton-Perkins. "Mulching with Wool: Opportunities to Increase Production and Plant Viability against Pest Damage While Creating New Regional Markets for Kempy (Unsalable) Wool." SARE Outreach. USDA. Accessed January 3, 2021. https://projects.sare.org/sare_project/fnc10-797/.

Oelke. E. A., E. S. Oplinger, and M. A. Brinkman. "Triticale." Corn Agronomy. University of Wisconsin-Madison. November, 1989. http://corn.agronomy.wisc.edu/Crops/Triticale.aspx.

Ogle, D., L. St. John, and D. Tilley. "Plant Guide: Yellow Sweetclover (*Melilotus officinalis* (L.) Lam. and White Sweetclover (*M. alba* Medik)." USDA Natural Resources Conservation Service, 2008. https://citeseerx.ist.psu.edu/viewdoc/download?doi=10.1.1.370.2589&rep=rep1&type=pdf.

O'Hara, Bryan. *No-Till Intensive Vegetable Culture: Pesticide-Free Methods for Restoring Soil and Growing Nutrient-Rich, High-Yielding Crops*. White River Junction, VT: Chelsea Green, 2020.

Oplinger, E. S., E. A. Oelke, M. A. Brinkman, and K. A. Kelling. "Buckwheat." Alternative Field Crops Manual. University of Wisconsin Extension. Updated January 13, 2021. https://hort.purdue.edu/newcrop/afcm/buckwheat.html.

Oregon State University. "Berseem Clover." Forage Information System. Updated April 15, 2018. https://forages.oregonstate.edu/forages/berseem-clover.

Palmer, Nigel. *The Regenerative Grower's Guide to Garden Amendments: Using Locally Sourced Materials to Make Mineral and Biological Extracts and Ferments*. White River Junction, VT: Chelsea Green, 2020.

Pavek, Pamela L. S. "Plant Guide: Pea (*Pisum sativum* L.)" USDA Natural Resources Conservation Service. Updated June 2012. https://plants.usda.gov/plantguide/pdf/pg_pisa6.pdf.

Pavlis, Robert. *Soil Science for Gardeners: Working with Nature to Build Soil Health*. Gabriola Island, BC: New Society, 2020.

———. "Should Collected Seed Be Stored in the Fridge or Freezer?" Garden Myths (website). Accessed January 3, 2021. https://www.gardenmyths.com/storing-collected-seed-fridge-or-freezer/.

Peigné, Joséphine, and Philippe Girardin. "Environmental Impacts of Farm-Scale Composting Practices." *Water, Air, and Soil Pollution* 153, (2004): 45–68. https://doi.org/10.1023/B:WATE.0000019932.04020.b6.

Perkins, Richard. *Regenerative Agriculture: A Practical Whole Systems Guide to Making Small Farms Work*. Self-published, 2019.

Picasso, Valentín D. "Diversity, Productivity, and Stability in Perennial Polycultures Used for Grain, Forage, and Biomass Production." PhD diss., Iowa State Universiy, 2008. ProQuest (UMI 3291999. https://www.researchgate.net/publication/36710330_Diversity_productivity_and_stability_in_perennial_polycultures_used_for_grain_forage_and_biomass_production.

Picasso, Valentín D., E. Charles Brummer, Matt Liebman, Philip M. Dixon, and Brian Wilsey. "Crop Species Diversity Affects Productivity and Weed Suppression in Perennial Polycultures under Two Management Strategies." *Crop Science* 48, no. 1 (2008): 331–42. https://doi.org/10.2135/cropsci2007.04.0225.

Poelman, Erik H., Si-Jun Zheng, Zhao Zhang, Nanda M. Heemskerk, Anne-Marie Cortesero, and Marcel Dicke. "Parasitoid-Specific Induction of Plant Responses to Parasitized Herbivores Affects Colonization by Subsequent Herbivores." *Proceedings of the National Academy of Sciences* 108, no. 49 (December 2011): 19647–52. https://doi.org/10.1073/pnas.1110748108.

Porter, Carly. "Why Do Earthworms Surface After Rain?" *Scientific American*, April 14, 2011. https://www.scientificamerican.com/article/why-earthworms-surface-after-rain/.

Postma, Johannes A., and Jonathan P. Lynch. "Complementarity in Root Architecture for Nutrient Uptake in Ancient Maize/Bean and Maize/Bean/Squash Polycultures." *Annals of Botany* 110, no. 2 (July 2012): 521–34. https://doi.org/10.1093/aob/mcs082.

Reddy, Rohini. *Cho's Global Natural Farming*. South Asia Rural Reconstruction Association, 2011.

Rees, Jenny, Roger Elmore, Chris Proctor, Steve Melvin, and Michael Sindelar. "Interseeding Cover Crops into Corn or Soybean." University of Nebraska Institute of Agriculture and Natural Resources. May 23, 2019. https://cropwatch.unl.edu/2019/interseeding-cover-crops-corn-or-soybean.

Sanchez, Elsa "Effect of Transplant Age on Yield." Vegetable Crops Online Resources. Rutgers University. Originally published in *The Vegetable & Small Fruit Gazette*, Penn State Cooperative Extension, June 2011. https://nj-vegetable-crops-online-resources.rutgers.edu/2015/04/effect-of-transplant-age-on-yield/.

Schuh, Marissa, and Will Jaquinde. "Impact of Soil Blocks on Yield and Earliness of Six Tomato Varieties." Michigan State University Extension. April 3, 2019. https://www.canr.msu.edu/news/impact-of-soil-blocks-on-yield-and-earliness-of-six-tomato-varieties.

Sheahan, C. M. "Plant Guide: Japanese Millet (*Echinochloa esculenta*)." USDA Natural Resources Conservation Service. Updated September 2014. https://plants.usda.gov/plantguide/pdf/pg_eces.pdf.

Sherman, Rhonda. *The Worm Farmer's Handbook: Mid- to Large-Scale Vermicomposting for Farms, Businesses, Municipalities, Schools, and Institutions*. White River Junction, VT: Chelsea Green, 2018.

Silva, Erin, Jed Colquhoun, Anne Pfeiffer, and Ruth McNair. "Living Mulch Suppresses Weeds and Yields in Organic Vegetable Plots (CIAS Research Brief #100)." Center for Integrated Agricultural Systems, UW-Madison. September 2018. https://www.cias.wisc.edu/living-mulch-suppresses-weeds-and-yields-in-organic-vegetable-plots-cias-research-brief-100/.

Stika, Jon. *A Soil Owner's Manual: How to Restore and Maintain Soil Health*. Self-published, CreateSpace, 2016.

Sulvai, Fraide, Beni Jequicene Mussengue Chaúque, and Domingos Lusitâneo Pier Macuvele. "Intercropping of Lettuce and Onion Controls Caterpillar Thread, *Agrotis ípsilon* Major Insect Pest of Lettuce," *Chemical and Biological Technologices in Agriculture* 3 (2016): article 28, https://doi.org/10.1186/s40538-016-0079-z.

UC-Davis Drought Management. "Crop Rooting Depth." Originally published in Chapter 11, "Sprinkler Irrigation," Section 15, Natural Resources Conservation Service National Engineering Handbook. http://ucmanagedrought.ucdavis.edu/Agriculture/Irrigation_Scheduling/Evapotranspiration_Scheduling_ET/Frequency_of_Irrigation/Crop_Rooting_Depth/.

University of Nebraska-Lincoln Institute of Agriculture and Natural Resources. "Relay Cropping Offers Economic Benefits and Reduces Nitrate Leaching from Soils." September 19, 2003. https://cropwatch.unl.edu/wheat/relaycrop.

Warren, Nicholas D., Richard G. Smith, and Rebecca G. Sideman. "Effects of Living Mulch and Fertilizer on the Performance of Broccoli in Plasticulture." *HortScience* 50, no. 2 (2015): 218–24. https://doi.org/10.21273/HORTSCI.50.2.218.

Witting-Bissinger, Brooke, D. B. Orr, and H. M. Linker. "Effects of Floral Resources on Fitness of the Parasitoids *Trichogramma exiguum* (Hymenoptera: Trichogrammatidae) and *Cotesia congregata* (Hymenoptera: Braconidae)." *Biological Control* 47, no. 2. (2008): 180–86. https://doi.org/10.1016/j.biocontrol.2008.07.013.

Young-Mathews, Annie. "Plant Guide: Crimson Clover (*Trifolium incarnatum*)." USDA Natural Resources Conservation Service. Updated February 2013. https://plants.usda.gov/plantguide/pdf/pg_trin3.pdf.

INDEX

Note: Page numbers in *italics* refer to figures and photographs. Page numbers followed by *t* refer to tables.

17 essential nutrients, 22

A

Ace of Spades Farm, 102
activated biochar, 40
Advancing Eco Agriculture, 239
Albrecht soil test method, 35–36
alfalfa meal, 143, *161*, 190, 221
allelopathy, 12, 183
allium family, interplanting benefits, 182
  See also specific types
alyssum (*Lobularia maritima*), cover crop use and termination, 245–46
amending the soil. *See* soil amendments
ammonia, in poultry manure, 64
animal products, in compost, 63
animals
  grazing of cover crops, 45, 153
  in the no-till market farm, *x*
  tillage by, 44–45, *44*
annual flax (*Linum usitatissimum*), cover crop use and termination, 242
aphids, cherry tomato concerns, 232, 239
Appropriate Technology Transfer for Rural Areas (ATTRA), 78
arid climates. *See* dry climates
Armour, Jay, 119
Armour, Polly, 119
arugula
  critical period of competition, 247
  in-depth profile of, 201–5, *203*, *204*
  interplanting strategies, 180, 204, 247
  shade for, 16
A:shiwi (Zuni) people, 54
Assawaga Farm, 76, 161
Asteraceae (daisy family), critical period and pairings, 253
ATP production, 11
Atthowe, Helen, 77, 101, 124
ATTRA (Appropriate Technology Transfer for Rural Areas), 78
Austrian winter peas (*Pisum sativum*)
  cover crop use and termination, 243–44
  for mulch-in-place living pathways, 126
  nitrogen fixation, *149*

B

baby beets, 229
baby carrots, *196*, 197, 199
Back to Eden (film), 80
bags
  for arugula, 205
  for lettuce, 216, 217–18
balers, for walk-behind tractors, 110
bare soil, degradation from, 23
bark mulch, 82, 118–19
barley (*Hordeum vulgare*), cover crop use and termination, 242
barrels, crimping with, 155
Baruc, Ricky, 79, 105
Base Cation Saturation Ratio (BCSR) system, 35–36, 141
basil, critical period and suggested pairings, 247
BCS walk-behind tractors, 30, 47, 107, *108*
beans
  critical period of competition, 247
  interplanting strategies, *180*, 247
  in the Three Sisters combination, 174, 184
  See also legumes; *specific types*
bed flips. *See* turning over beds
bed preparation
  arugula, 201
  beets, 226
  carrots, 195–96
  cherry tomatoes, 230
  garlic, 206

bed preparation (*continued*)
lettuce, 213–14
for relay cropping, 190
sweet potatoes, 221
without tillage, 159–164, *160–61, 163–64*
bed rakes, 159
bed rollers, 155, 162–63
beds
designing, 47–53, *49, 51*
establishing, 53–56, *54–56, 67*
lasagna beds, 55, *55*, 104, *104*
lowered beds, 54, *54*
raised beds, 52–53, 55
splitting crops in, *178*, 180
See also turning over beds
bees, in mulch, 73
beets
critical period of competition, 247–48
in-depth profile of, 226–230, *227, 228*
high tunnel cultivation of, 16, *16*
interplanting strategies, 179, 180, *180*, 248
relay cropping with, 190
rhizosheaths on roots of, *147*
beneficial organisms
cardboard mulch considerations, 73
interplanting for, 175–78, *176, 177*, 187
perennials for, 186–87
berseem clover (*Trifolium alexandrinum*), cover crop use and termination, 244
Berta flail mowers, 107
billboards, occultation with, 98
bioassays (growth tests), 68
biochar, activated, 40
biodegradable plastic mulches (biofilms), 101
biostrip till, as term, 124
See also strip tillage systems
Black people, land stewardship by, 5
black polyethylene sheets, 97
See also silage tarps
black walnut, growth inhibitors from, 82
blister beetles, 226
boards, crimping with, *154*
Bokashi composting, 144
Bolero carrots, 195, 197
book resources, 256–57
bottom watering, 168
braconid wasps, *177*
brassicas
allelopathic properties, 183
cover crop use and termination, 245–46
nurse crops for, 191
See also specific types
breaking new ground, 33–56
amending the soil, 40
animal tillage, 44–45, *44*
checking compaction, 37–40
clearing a forest, 41
designing permanent beds, 47–53, *49, 51*
establishing permanent beds, 53–56, *54–56*
managing compaction, 40–42, 46
never-till approach, 42–44, *43*
site selection, 33–35
soil tests, 35–37, *36*
transitioning to no-till, 45–47, *46*
breeding of new varieties, 194
Briceland Forest Farm, 104
Broadfork Farm, 70
broadforks
appropriate use of, 3–4, 26
managing soil compaction with, 41, 46, *118*
broadleaf herbicides, 68, 75
broccoli, critical period and suggested pairings, 248
Brown, Gabe, 153
bubbling, of lettuce, 214
buckwheat (*Gaopyrum esculentum*)
as cover crop, 149, *152*, 241
termination methods, 241
Buena Vista Gardens, 205
The Buffalo Seed Company, 194, 257
bush blades, for weed whackers, 102, 105, 157–58
bush hog mowers, for turning over beds, 110

C

cabbage, critical period and suggested pairings, 248
Cannady, Zach, 197
capillary mats, 168
carbonaceous materials
in inoculating composts, 61–62
in lasagna beds, 55, 56, 104
in mulching composts, 65, *66*
in nutritional composts, 65–66
carbon cycle
benefits of leaving roots in place, 91–92
role of photosynthesis in, *10*, 13, 17
carbon dioxide
importing, 18
role in photosynthesis, *10*, 11, 15, 17–18
carbon fixation, 17

carbon sequestration
  peat bog benefits, 84
  role of photosynthesis in, 13
carbon to nitrogen ratio (C:N ratio), 60, 65
cardboard mulch
  in cover crop termination, 157
  in the never-till approach, 43
  overview, 78–79, *78*
  in pathways, 117–18
  when establishing permanent beds, 55
  when transitioning to no-till, 45, *46*
  when turning over beds, 101–2
Carpenter, Alex, 76, 161
carrots
  baby carrots, *196*, 197, 199
  critical period of competition, 248
  in-depth profile of, 195–201, *196*, *198*, *200*
  interplanting strategies, 172, *173*, 174, 248
  interplanting with, 4
  as nurse crop, 191
  relay cropping with, 188, 190, 199
caterpillars, parasitoids of, 178
caterpillar tunnels
  controlling growing conditions in, 15
  orientation of beds in, 52
  sun protection with, 16
cation exchange capacity, 35
cats, for rodent control, 73
cauliflower, critical period and suggested pairings, 248
celery
  critical period and suggested pairings, 248–49
  interplanting strategies, 174, *175*
  relay cropping with, 188
cell trays, 167–68
Cercospora leaf spot, 226
cereal rye (*Secale cereale*)
  cover crop use and termination, *38*, 242
  in fertility management, 149
  for mulch-in-place living pathways, 126
  in the never-till approach, 43–44
cherry tomatoes, in-depth profile of, 230–39, *231*, *234–37*
chickens
  for animal tillage, 44–45, *44*
  compost from manure of, 64
Chickshaws, *44*, 45
chlorophyll, 20
chloroplasts, 11
Cho, Han-Kyu (Master Cho), 61, 138
Choate, Heidi, 55
Cho's Global Natural Farming (Reddy), 138, 139
Ciancioso, Stephen, 205
cilantro (*Coriandrum sativum*)
  cover crop use and termination, 241
  interplanting for beneficial organisms, 177, 178, 232
  as nurse crop, 191
clay-dominated soils
  amending with compost, 137, 140
  managing soil compaction, 41
clovers
  cover crop use and termination, 244, 245
  fresh mulch of, 77
  in living pathways, *121*, 122, 123, 126
C:N ratio (carbon to nitrogen ratio), 60, 65
coconut coir, as peat moss substitute, 84
Coleman, Clara, 171
Coleman, Eliot, 171, 256
common vetch (*Vicia sativa*), cover crop use and termination, 245
Community Machinery, 70, 257
compacted soil
  bed width considerations, 50
  checking for, 37–40
  effects of turning over beds with tillage, 89–90
  managing, 40–42, 46
  in pathways, *118*
  from pooled water on silage tarps, 99
  preparation for relay cropping, 190
  seed germination needs, 162
  soil structure concerns, 137
  from tillage, 25
  from wheel hoes, 111
competition, critical period of. *See* critical period of competition
compost, 59–70
  adding with soil amendments, 40
  animal products in, 63
  beds of, 43
  beneficial microorganisms in, *61*, *64*
  Bokashi composting, 144
  contaminants in, 68–69
  deep compost mulch system, 50, 66, 69–70, *69*
  defined, 60
  fertilizing composts, 64
  finding, 65, 70
  for inoculating biochar, 40

compost (*continued*)
  inoculating composts, 60–64, *62*, *63*, 138, 139, 143, 146, 190
  mulching composts, 65–66, *66*, 83, 119, 160, 197
  municipal composts, 68
  in the never-till approach, 43–44
  nutritional composts, 65, 66
  overview, *x*, 59–60, *59*
  as pathway mulch, 119
  risks with, 67–69
  setting up new beds with, 55, *67*
  shallow compost mulch system, 2, 38–39
  temperature of, 62, 63, 64
compost spreaders, 55, 70
compost teas and extracts
  in fertility management, 145–46, *145*
  for inoculating biochar, 40
  nourishing transplants with, 20, 169
containers, for plant growing, 167–68
conventional agriculture, environmental harm from, 5–6
corn
  critical period of competition, 249
  interplanting strategies, 179, 249
  in the Three Sisters combination, 174, 184
costs
  mulch paper, 79
  straw, 72
  synthetic mulches, 85
cover crops
  bringing up soil nutrients using, 22
  considerations for using raised beds, 53
  covering with peat moss after sowing, *83*, 84
  crimping of, 86, *87*
  defined, 86
  in fertility management, 142, 148–159
  grazing with sheep and goats, 45
  large-scale no-till cover cropping, 153
  as mulch, 39, 55–56, 86–88, *87*, *88*
  mulch-in-place living pathways, 125–26, *125*
  in the never-till approach, 43–44
  in the no-till market farm, *x*
  perennial, 150, *150*
  soil organic matter from, *14*
  tillage radishes, 42
  timing of, 56, 183
  uses of, 86–88, *87*, *88*, 241–46
  when establishing permanent beds, 55–56, *56*
  See also termination of cover crops
covering the soil
  as farming principle, 1–5
  nature as the model for, 31–32
  See also cover crops; mulch
cowpea (*Vigna unguiculata*), cover crop use and termination, 158, 244
Crabtree, Hannah
  bed length decision, 51
  fertility management program, 142
  initial crop failures, 1
  lettuce sign, *217*
  managing soil compaction, 40
  marketing system, 193
  move to new farm, 240
  shallow compost mulch system use, 2
  tomato training, 235
  trials of no-till methods, 38, 45, 89
Crickmore, Conor, 171, 256
Crickmore, Kate, 171
crimp and solarize method of cover crop termination, 156–57, *156*, *157*
crimp and tarp method of cover crop termination, 153–55, *154*, *155*
crimping, of cover crops, 86, *88*
  See also roller crimpers
crimp/mow and mulch method of cover crop termination, 157
crimson clover (*Trifolium incarnatum*)
  cover crop use and termination, *38*, 244
  for mulch-in-place living pathways, 126
Crispin, Kasey, 197
critical period of competition
  crop listings, 247–253
  overview, 181–82, *181*
Cropp, Jan-Hendrik, 77
crop spacing
  arugula, 202
  beets, 228
  carrots, 197
  cherry tomatoes, 233
  dry farming, 133
  garlic, 207
  lettuce, 215
  sweet potatoes, 223
cucumbers
  critical period of competition, 249
  interplanting strategies, 179, 249
  lowering and leaning training method, 235
cultivating shoe attachments, 128

curing, sweet potatoes, 224–25, *224*, 225

D

daikon radish (*Raphanus sativus* var. *niger*), 42
daisy family (Asteraceae), critical period and pairings, 253
Dawling, Pam, 190
deep compost mulch system
  mulching composts in, 66
  principles of, 69–70, *69*
  stepping on beds with, 50
deep-rooted crops list, 184
denitrification, 26
desertification, from destructive soil practices, 23
Desroches, Maude-Hélène, 116, 171
Dilley, Adam, 216
Dilley, Amanda, 216
Dirt Hugger, 167
disease control, interplanting benefits, 179, 182–83
disturbing the soil, minimizing, 1–5
double leader tomato plants, 233, 235, 236
Dowding, Charles, 50, 114, 177, 255, 256
drainage
  considerations for using raised beds, 53
  indicators of, 132
  orientation of beds for, 52
  site selection considerations, 34
dry climates
  considerations for using raised beds, 53
  deep pathways for, 114
  lowered beds in, 54, *54*
dry farming, 133, *133*
Dust Bowl, 23
Dyck, Bryan, 70

E

Earth Care Farm, 167
EarthWay seeders
  adding soil amendments using, 161, *161*, 230
  for arugula, 202
  for beets, 227–28
  for carrots, 197
earthworms
  affinity for cardboard, 118
  as indicator of living soil, *36*, *136*
  vermicast from, 61
edgers, for path management, *123*, 124
eggplant, critical period and suggested pairings, 248–49
Ekins, Alex, 102
England, Josh, 194
equipment. *See* tools and equipment
everbed farming. *See* relay cropping

F

fallowing periods, 148
Farmers Friend, 204, 257
farmers markets, selling at, 193–94
farming principles, 1–5
fava beans, critical period and suggested pairings, 248–49
feather meal, 95, 143, *161*, 190
fennel, critical period and suggested pairings, 248–49
La Ferme des Quatre-Temps, 171
fermented plant juice (FPH), 139
fertility management, 131–164
  bed preparation without tillage, 159–164, *160–61*, *163–64*
  cover cropping overview, 148–150, *149*
  cover crop termination methods, 150–59, *152*, *154–57*
  designing a fertility program, 142–48, *145*, *147*
  Korean Natural Farming, 138–39, *138*, *139*
  large-scale no-till cover cropping, 153
  living plant roots, 135, 137, 141–42
  perennial cover cropping, 150, *150*
  six factors in, 131
  soil biology, 140–41, *140*
  soil organic matter, 134–35, *134*, 140
  soil permeability, 137, 140, 142–43
  soil structure, 135–37, *136*, 142–43
  water, 132–34, *133*
fertilizing composts, 64
  See also compost
field peas (*Pisum sativum*)
  cover crop use and termination, 244
  for mulch-in-place living pathways, *125*, 126
flail mower collectors, 77, 109–10
flail mowers
  crimping with, 154
  for living pathways, 122
  for shredding leaves as mulch, 82
  for terminating cover crops, 87
  for turning over beds, 107–8, *108*, 109, *109*
flame termination of cover crops, 159
flame weeding, 196, 197

Flat Creek Farm, 218
flax (*Linum usitatissimum*), cover crop use and termination, 242
flea beetles, 202, 205
flipping beds. *See* turning over beds
Florida weave pattern, 106, 235, *235*
flowers
  interplanting strategies, 175–78, *176*, *177*
  native insect specialists, 187
  in the no-till market farm, *x*
food safety concerns, 122, 146
Food Safety Modernization Act, 63
food web, soil, 140–41
forage radish (*Raphanus sativus* var. *niger*), 42
forests, clearing, 41
Fortier, Jean-Martin, 96, 116, 134, 171, 256
Fort Vee potting mix, *167*
Four Seasons Farm, 171
Four Winds Farm, 119
FPH (fermented plant juice), 139
fresh hay mulch, 77–78
  See also mulch
Frith Farm, 153, 177
Fulmer, Daniel, 45
Fulmer, Hana, 45
fungi
  carbon dioxide respiration, 17
  as indicator of living soil, *25*
  mycorrhizal, 24
  no-till benefits for, *31*
  saprophytic, 113–14, *114*
  in spoiled hay, 77
  tillage effects on, 24

G

gaps, interplanting in, 172–74, *173*
garden shears, for turning over beds, 106
garden staples, 85
garlic
  cardboard mulch for, 101–2, *102*
  critical period and suggested pairings, 249–250
  in-depth profile of, 205–12, *208*, *209*, *211*
Gautschi, Paul, 80
germination, conditions for, 162, *163*, 166, 168
germination chambers, 168
ginger, critical period and suggested pairings, 250
goats, for grazing cover crops, 45
gopher control, 73
grafting, of tomatoes, 232
grass clippings, for mulch, 77–78, *200*
  See also mulch
grasses, in pathways, 122–24, *123*
grass seed, covering with peat moss, 83
green garlic, 206, 207, *208*
  See also garlic
greenhouses
  carbon dioxide enrichment, 18
  controlling growing conditions in, 15
  curing sweet potatoes in, 224, 225
  growing conditions in, 168–69
green onions
  critical period of competition, 251
  interplanting strategies, 179, 180, *180*, 182, *183*, 251
  as nurse crop, 191
  relay cropping with, 188, *189*
greens
  beets, 226
  sweet potatoes, 225
  See also lettuce
Gridder tool, 207
ground hornets, in mulch, 73
Growers Live (YouTube channel), 2
growing season, lengthening with no-till practices, 29
growing space
  bed size considerations, 47–48
  for cover crops, 148–49
  maximizing through interplanting, 172
grow-in-place mulch (cover crop mulch), 39, 55–56, 86–88, *87*, *88*
growth tests (bioassays), 68
guard cells, 18
gypsum, adding to clay soils, 140

H

Haartman, Daniel, 70
Habib, Deb, 79, 105
habitat for wildlife, 32, *32*
hairy vetch (*Vicia villosa*), cover crop use and termination, *38*, 245
Halley, Logan, *104*
Haney Test, 141
Hansch, Corinne, 84
hardening off, of transplants, 168–69
hardneck vs. softneck garlic, 205–6
  See also garlic

harrows
  crimping with, 154–55, *155*
  for seedbed preparation, 163
Hartman, Ben, 204
Harvard University life satisfaction study, 30
harvesting
  arugula, 202, 204, *204*
  beets, *227*, 228–29, *228*
  carrots, *196*, 197, 199
  cherry tomatoes, 236, 238
  garlic, 207, 210
  greater focus on, with no-till practices, 29–30
  lettuce, 216–17, *216*
  sweet potatoes, 223–24
haylage mulch, 77–78
  See also mulch
hay mulch, *x*, 74–77, *75*
heat termination of cover crops, 159
hedgerows
  for beneficial organisms, 187
  mushroom cultivation in, 115
  in the no-till market farm, *x*
hemp, sunn (*Crotalaria juncea*)
  cover crop use and termination, 158, 245
  for mulch-in-place living pathways, 126
herbicides
  compost contaminated with, 68
  mulch contaminated with, 75
  straw contaminated with, 74
herbs, interplanting strategies, 177, 178
high tunnels
  bed length, 51
  controlling growing conditions in, 15
  landscape fabric use in, 85
  path width, 50
  solarization using plastic from, 157
  sun protection with, 16, *16*
  using weed whackers in, *103*, 105
Hinke, Troy, 61
hoes
  for cultivating pathways, 127, 128
  for turning over beds, 110–11
hornworms, *177*, 178, 232
Houtz, Robert L., 17
Hozon siphons, 161
hugulkultur beds, 104
humic acid, 143, 190
humidity
  for curing sweet potatoes, 224
  seed germination needs, 166
hyphae, fungal, 24

I

indigenous microorganism (IMO) collection, 138–39, *138*, *139*
indigenous peoples
  interplanting practices, 171
  land stewardship by, 5, 22
  lowered beds for dry climates, 54, *54*
Ingham, Elaine, 140, 141, 256
inoculating composts
  in compost teas and extracts, 146
  creating, 60–64, *62*, *63*
  in fertility management, 138, 139, 143, 190
  See also compost
insects, beneficial
  cardboard mulch considerations, 73
  interplanting for, 175–78, *176*, *177*, 187
  perennials for, 186–87
interplanting
  arugula, 204
  beets, 229
  beginner combinations, 172, 179–180, *180*
  for biocontrol, 182–83, *183*
  carrots, 4, 199
  cherry tomatoes, 238
  critical period of competition, 181–82, *181*
  crop listings, 247–253
  encouraging beneficials, 175–78, *176*, *177*
  by filling in gaps, 172–74, *173*
  by filling unused space, 172
  garlic, 210, *211*, 212
  lettuce, 217, *219*
  maximizing profits by, *x*, 178–79, *178*
  nurse crops, 191–92
  overview, 169, 171–72
  perennial additions, 186–87, *186*
  relay cropping, 183, 187–190, *189*
  root complementarity, 184–86, *185*
  structural benefits of, 174–75, *175*
  sun protection with, 17
  sweet potatoes, 225
irrigation
  beds next to living pathways, 124
  bottom watering, 168
  for cardboard mulching, 79
  fertility management considerations, 132, 134
  site selection considerations, 34

irrigation (*continued*)
  when establishing permanent beds, 56
  See also water

J

Jadrnicek, Shawn, 153
Jang seeders, 153, 197, 202, *203*, 227–28
Japanese millet (*Echinochloa esculenta*), cover crop use and termination, 242
Japanese radish (*Raphanus sativus* var. *niger*), 42
Les Jardins de la Grelinette, *116*, 171
Jared's Real Food, 104, 106, 172
Johnny's Selected Seeds, 202, 213, 214, 257
Jones, Christine, 146, 177, 184, 256
Jones, Shannon, 70
JP-1 seeders. *See* Jang seeders
Juglans spp., growth inhibitors from, 82
juglone, 82

K

Kaiser, Elizabeth, 70
Kaiser, Paul, 70
kale
  bed width considerations, *49*
  critical period and suggested pairings, 250
  relay cropping with, 188
keeping the soil planted, as farming principle, 1–5
kelp, in fertility management, 143, 190
Kempf, John, 239, 255
king stropharia mushroom (*Stropharia rugosoannulata*), 115, *115*
knives, for turning over beds, 106–7, *107*
Korean Natural Farming, 61, 138–39, *138*, *139*, 144
Kost, Nancy, 194

L

lacewings, for pest control, 232, 239
lacy phacelia (*Phacelia tanacetifolia*), cover crop use and termination, 241–42
ladybugs, for pest control, 232, 239
landscape fabrics
  mulching with, 85–86, *85*
  occultation with, 99–100, *100*
  in pathways, 120
Landzie Compost and Peat Moss Spreaders, *83*, 84, 196, 221
lasagna beds
  for new beds, 55, *55*, 104, *104*
  for sweet potatoes, 221
Latdekwi:we (waffle gardening), 54, *54*
lateral growth crops list, 186
lawn staples, 85
The Lean Farm Guide to Growing Vegetables (Hartman), 204
leaves and leaf mold mulches, 82–83
leeks, critical period and suggested pairings, 250
leggy seedlings, 16, 168
legumes
  as cover crop, 149
  in living pathways, 123
  nitrogen fixation with, 146
  as nurse crop, 191–92
  See also beans; *specific types*
Lein, Susana, 86
length of permanent beds, 50–51
Leon, Matthew, 84
lessons learned
  arugula, 205
  bed sizes, 47
  beets, 229–230
  carrots, 200–201
  cherry tomatoes, 239
  compacted soil, 40, 46
  garlic, 181, 212
  lettuce, 218, 220
  perennial cover crops, 150
  sweet potatoes, 225
  transitioning to no-till, 45–47
  turning over beds, 89–90
lettuce (*Lactuca sativa*)
  bed width considerations, *49*
  cover crop use and termination, 242
  critical period of competition, 250
  in-depth profile of, 212–220, *213*, *215–17*, *219*
  interplanting strategies, 172, *173*, 174, 178, 179, 180, *180*, 182, *183*, 217, 250
  leaving roots in place, *92*
  relay cropping with, 188
  root complementarity with tomatoes, 184
  shade for, 16
  transplants of, *166*
ley fields, fresh hay from, 77
living pathways
  management of, 120–26, *121*, *123*, *125*
  in the no-till market farm, *x*
  between okra beds, *4*
  width of, 50
Living Roots Compost Tea, 61

living soil, as term, 12
  See also specific practices
long-season occultation, 42–43, *43*
loppers, for turning over beds, 106
Love, Jennie, 123–24
Love 'n Fresh Flowers, 123
Lovin' Mama Farm, 84
lowered beds, 54, *54*
lowering and leaning, 234–35, 236, *237*

M

machetes, for turning over beds, 106–7
magazine recommendations, 255
magnesium, role in photosynthesis, 20
manganese, role in photosynthesis, 20
manures
  fertilizing composts from, 64
  nutritional composts from, 65
  phosphorus in, 67
manure spreaders, for fresh hay mulch, 77
marigolds, interplanting benefits, 182
The Market Gardener (Fortier), 96, 171, 256
marketing
  arugula, 205
  beets, 228, 229
  carrots, *198*, 199
  cherry tomatoes, 238
  garlic, 210
  lettuce, 217–18, *217*
  overview, 193–94
  sweet potatoes, 225
marking plots, 53–55
Mays, Daniel, 153, 177, 256
microBIOMETER soil tests, 141
micronutrients
  checking for deficiencies in, 36, 144
  role in photosynthesis, 20, 22
microplastics, from silage tarps, 98
milk stage of cereal grains, 151
millet, Japanese (*Echinochloa esculenta*), cover crop use and termination, 242
millet, pearl (*Pennisetum americanum*), cover crop use and termination, 243
mineral balancing, 36, 95, 144
mineralization of soil organic matter, 147–48
modified relay intercropping (MRI), 187
  See also relay cropping
moisture. *See* water
mowing
  for living pathways, 120, 122
  for turning over beds, 107–10, *108*, *109*
Moyer, Jeff, 86
Muir lettuce, 212, *213*
mulch, 71–88
  bark, 82, 118–19
  benefits of, 28–29, *28*, 71–72
  cardboard, 43, 45, *46*, 55, 78–79, *78*, 101–2, 117–18, 157
  cover crops, 39, 55–56, 86–88, *87*, *88*
  deep compost mulch system, 50, 66, 69–70, *69*
  fresh green materials, 77–78
  hay, 74–77, *75*
  with large-scale no-till cover cropping, 153
  leaves and leaf mold, 82–83
  nature as the model for, 31–32, 71
  in the never-till approach, 43–44
  paper, 43, 45, 55, 78–79, *78*, 101–2, 117–18, 157, 164
  peat moss, 83–84, *83*
  pests in, 73
  sawdust, 82, 118–19
  shallow compost mulch system, 2, 38–39
  straw, 72–74, 76, *76*
  synthetic, 85–86, *85*, 119–120
  when establishing permanent beds, 55–56
  when transitioning to no-till, 45, *46*
  when turning over beds, 95
  woody materials, 80–82, *81*
mulching composts
  applying soil amendments beneath, 160
  for carrots, 197
  creating, 65–66, *66*
  in pathways, 119
  peat moss in, 83, 119
mulching mowers. *See* flail mowers
mulch-in-place living pathways, 125–26, *125*
mulch paper, *78*
  See also paper mulch
multicropping. *See* interplanting
mung beans (*Vigna radiata*), cover crop use and termination, 244
municipal composts, 68
municipal water, chemicals in, 132
mushroom cultivation
  in beds, *140*
  in pathways, 115, *115*
mustard (*Guillenia flavescens*), cover crop use and termination, 246

mycorrhizae, 24, 191
mycorrhizal fungi, 24

N

NADPH production, 11
Napa cabbage, relay cropping with, 188, *189*
National Organic Program (NOP)
 compost standards, 63, 68
 listed soil amendments, 40
 silage tarp standards, 100
native insects, perennials for, 186–87
Natural Resources Conservation Service (NRCS), 23, 137
nature as the model for no-till practices, 31–32
NatureFlex bags, 205
Neversink Farm, 171
Neversink Farm Tools, 207, 257
never-till approach, 42–44, *43*
New England Compost, 167
new ground. *See* breaking new ground
The New Organic Grower (Coleman), 171, 256
nightshades
 hay mulch for, *75*
 interplanting strategies, 177, *177*, 179–180, 204, 217, 248
 See also specific types
nitrates
 in contaminated groundwater, 132
 plant-available nitrogen in, 64, 147
nitrogen
 compacted soil effects on, 26
 in compost, 64, 67
 fertility management for, 95, 146–48, *147*
 meals as source of, 143
 reduced availability from decomposing woody materials, 80, 82
 role in photosynthesis, 20
 in soil organic matter, 14
nitrogen fixation, 146–47, *149*, 191–92
nitrogen-fixing bacteria, 146–47
nitrogenous materials
 in inoculating composts, 62
 in lasagna beds, 55, 104
 in mulches, 79, 80
no-dig method. *See* deep compost mulch system
nodules, in nitrogen fixation, 146, *149*
no-mulch methods, 170–71, *170*
no mulch pathways, 127–28, *127*
NOP. *See* National Organic Program (NOP)
no-till agriculture
 appropriate disturbances of the soil, 27
 elements of, x
 farming principles, 1–5
 nature as the model for, 31–32
 practical reasons not to till, 27–30, *28*
 reasonable use of no-till principles, 3–4
 transitioning to, 45–47, *46*
 See also specific practices
No-Till Flowers Podcast, 123
No-Till Growers (organization), 2, 255, 256
No-Till Intensive Vegetable Culture (O'Hara), 111, 161, 256
The No-Till Market Garden Podcast, 2, 45, 190
The No-Till Organic Vegetable Farm (Mays), 153, 256
NRCS (Natural Resources Conservation Service), 23, 137
nurse crops, 191–92
nutrient deficiencies
 Korean Natural Farming recipes for ameliorating, 139
 mineral balancing for, 36, 95, 144
 soil testing for, 35–37
nutrients
 17 essential nutrients, 22
 in compost, 67
 mycorrhizae networks for, 24
 replenishing when removing material, 78
 role in photosynthesis, 15, 20–22, *21*
nutritional composts, 65, 66

O

oats (*Avena sativa*)
 cover crop use and termination, 158, 242–43
 for mulch-in-place living pathways, 126
occultation
 for cover crop termination, 153–55
 long-season, 42–43, *43*
 solarization vs., 111, 112
 when creating pathways, 120
 when turning over beds, 96–102, *96*, *100*, *102*
O'Hara, Bryan, 111, 112, 161, 256
OHN (oriental herbal nutrient), 139
oilseed radish (*Raphanus sativus* var. *oleiformis*), as tillage radish, 42
okra
 critical period of competition, 250–51
 interplanting strategies, 179, 225, 250–51
 living pathways for, *4*, *121*

OMRI (Organics Materials Review Institute), 79
onions, bulb
  critical period of competition, 251
  interplanting strategies, 182, 251
onions, green. *See* green onions
online resources, 255–56, 257
organic agriculture
  cardboard and paper mulch requirements, 79
  compost teas and extracts use, 146
  plastic mulch use, 100, 101, 119
  slug control, 73
  soil amendment considerations, 40
  synthetic mulch use, 86
organic matter. *See* soil organic matter
Organics Materials Review Institute (OMRI), 79
oriental herbal nutrient (OHN), 139
orientation of beds, 51–52
oxidative stress, 16
Oxton Organics, 172
oxygen
  germination role, 162, 163
  nitrogen fixation role, 146, 147
  photosynthesis role, *10*, 11

P

paper mulch
  in cover crop termination, 157
  in the never-till approach, 43
  overview, 78–79, *78*
  in pathways, 117–18
  for soil temperature management, 164
  when establishing permanent beds, 55
  when transitioning to no-till, 45
  when turning over beds, 101–2
Paperpot Co, 202, 215, 257
Paper Pot Transplanters, 202, 215
parasitoids, 178
Parnes, Bob, 101
parsley
  as nurse crop, 191
  pairings for, 248
parsnip, critical period and pairings, 251
path management, 113–128
  living pathways, *x*, *4*, 50, 120–26, *121*, *123*, *125*
  mulches for, 113–120, *114–18*
  mushroom cultivation in paths, 115, *115*
  overview, 113
  unmulched paths, 127–28, *127*
pathogenic bacteria, in contaminated water, 132
pathways
  bed width considerations, 49–50
  width of, 50, 120, 122
  wood chips for, 80
  See also living pathways
Pavel's Garden, 167
pearl millet (*Pennisetum americanum*), cover crop use and termination, 243
peas
  as cover crop, 158
  critical period of competition, 251
  interplanting strategies, 251
  as nurse crop, 191–92
  See also Austrian winter peas (*Pisum sativum*); field peas (*Pisum sativum*)
peat moss mulch, 83–84, *83*
Peña, Brijette, 80
penetrometers, 38
peppers
  critical period of competition, 252
  interplanting strategies, 178, 179, *180*, 252
  transplants of, 233
percentage slope calculations, 35
perennials
  cover crops of, 150, *150*
  interplanting strategies, 186–87, *186*
  in the no-till market farm, *x*
Perkins, Evan, 55
permanent beds. *See* beds
permeability, soil, 137, 140, 142–43
pest control
  arugula, 202, 205
  beets, 226, 229–230
  cherry tomatoes, 232, 239
  interplanting benefits for, 179, 182–83, *183*
  lettuce, 214
  in mulch, 73
  sweet potatoes, 222
  wildlife for, 32
pH of soil, 35, 37
phosphorus, in compost, 67
photosynthesis
  feeding the soil via, 11–15, *14*
  five keys to, 15–22, *16*, *19*, *21*
  overview, 9–11, *10*
phthalates, 86, 98
physics balancing, 163, 171
phytochemicals/phytonutrients, creation of, 12
pigs, for animal tillage, 44

pinpoint seeders, 202
plant canopies, nature as the model for, 31
planting
greater focus on, with no-till practices, 29–30
managing soil compaction with, 41–42
replanting when turning over beds, 96
See also seeding; transplants
plastic mulches
mulching with, 85–86
occultation with, 100–101
in pathways, 119–120
polyculture. *See* interplanting
polyethylene black plastic sheets, 97
See also silage tarps
potassium, role in photosynthesis, 20
potatoes
critical period and suggested pairings, 252
relay cropping with, 188
potting mixes, 166–67, *167*
power harrows
crimping with, 154–55, *155*
for seedbed preparation, 163
precision depth rollers, 160, 163
Prema Farm, 197
pricing systems, 194
probes, soil, *37*, *134*
profitability, interplanting strategies for, 178–79, *178*
propagation houses, in the no-till market farm, *x*
ProtekNet, 205, 226
pruning, of tomato plants, 235–36
push mowers, for turning over beds, 108–9
PVC (polyvinyl chloride), in used billboards, 98

Q

quality of life benefits of no-till practices, 30
Quick-cut Greens Harvesters, 204, *204*

R

radishes
critical period of competition, 252
interplanting strategies, 179, 252
relay cropping with, 188, *189*, 190
tillage, 42
rainfall
orientation of beds considering, 52
pathway materials and, 114, 117
raised beds, 52–53, 55
rakes
for incorporating soil amendments, 159
for seedbed preparation, 163
for turning over beds, 108
Ramblin' Farmers, 104
ramial wood chips, 80, 116–17, *116*
See also wood chips
Raphanus sativus varieties, as tillage radishes, 42
reading recommendations, 255–57
rebar, for measuring soil compaction, 39
red clover (*Trifolium pratense*), cover crop use and termination, 244
Reddy, Rohini, 138, 139
relay cropping
arugula, 204
beets, 229
carrots, 199
cherry tomatoes, 238
defined, 183
principles of, 187–190, *187*
replanting
greater focus on, with no-till practices, 29–30
timing of, 96
See also turning over beds
resources, 2, 255–57
rhizobia, 146
rhizophagy cycle, 12, 141
rhizosheaths, *147*
rhizosphere
incorporating soil amendments in, 159–162, *160*, *161*
sampling for soil tests, 37
Ritchey, Kip, 76
road access, site selection considerations, 34
rocket. *See* arugula
Rodale Institute, 152
rodent control, 73
Rolett, Jackson, 2
roller crimpers
considerations for using raised beds, 53
for terminating cover crops, 151–52, *152*
uses of, 86, *88*
rollers, bed, 155, 162–63
root complementarity, 184–86, *185*
root exudates
biodiversity benefits, 175, 177
in living soil, 12, 13
nourishment of soil microorganisms, *21*

roots
  carbon in, 13–14
  in fertility management, 141–42, 143
  food web of, *21*
  leaving in place when turning over beds, 91–93, *92*, *93*
  mycorrhizal fungi on and in, 24
  soil organic matter from living plant roots, 135, 137
Rosato, Carl, 77, 101, 124
Rose Creek Farms, 99, 218
rotary plows
  for creating permanent beds, 47, 55
  for path management, 117
rotary tillers, appropriate use of, 26
row covers
  for beets, 226
  for carrots, 200–201
  for lettuce, 220
  soil temperature management with, 164
RuBisCo (ribulose-1,5-bisphosphate carboxylase-oxygenase), 17
Rudolph, Spencer, 48
rye cover crops. *See* cereal rye (*Secale cereale*)
ryegrass, in living pathways, *121*, 123
Ryler, Ray, 218

S

safety considerations
  food safety concerns, 122, 146
  greenhouse carbon dioxide enrichment, 18
Sage Hill Ranch Gardens, *48*
Salamander Springs Farm, 86
Salanova Foundation Collection lettuce mix, 212, 213
salt marsh hay, 76
  See also hay mulch
San Diego Seed Company, 80
sandy soils, improving, 140
saprophytic fungi, 113–14, *114*
SARE (Sustainable Agriculture Research and Education), 119
Sattin, Josh, 2, 80, 235, 255
sawdust mulch, 82, 118–19
scapes, garlic, 207, *209*
  See also garlic
Sclerotinia drop, 214
scurf (fungal disease), 222, *222*
scythes and scything, 105–6, *106*
seasonal considerations
  arugula, 205
  beets, 229
  carrots, 199–200
  cherry tomatoes, 238--39
  cover crop planting, 56, 183
  establishing permanent beds, 53
  garlic, 210, 212
  lettuce, 218
  raised bed adjustments, 52
  termination of cover crops, 88
secondary metabolites, 12
seedbeds, prepping, 162–64, *163*, *164*
seed company listings, 257
seeders
  for arugula, 202, *203*
  for beets, 227, 229
  for carrots, 197
  large-scale farm use, 153
  for side dressing crops, 161, *161*
  simultaneous crimping and planting, *88*
seeding
  arugula, 202, *203*
  beets, 226, 227–28
  carrots, 195, 197
  garlic, 206–7
  lettuce, 214–15
  moisture retention benefits of peat moss while, *83*, 84
  mulch-in-place living pathways, 126
  nurse crops, 192
  prepping seedbeds, 162–64, *163*, *164*
  in straw or hay mulch, 76
seedlings, leggy, 16, 168
seeds
  ensuring quality of, 165–66
  storage of, 166
Seeds of Solidarity Farm, 79, 104, 105
seed starting
  beets, 227–28
  cherry tomatoes, 232–33
  in greenhouses, 168–69
seed-starting mixes, 166–67, *167*
Sesbania spp.
  cover crop use and termination, 245
  for mulch-in-place living pathways, 126
17 essential nutrients, 22
shade
  for beets, 226

shade (*continued*)
methods of providing, 15, 16–17
from tall crops, 174, 175, *175*
shade avoidance mechanism. *See* critical period of competition
shade cloth
for lettuce, *215*, 220
in the no-till market farm, *x*, 15
shallow compost mulch system, 2, 38–39
See also mulching composts
shallow-rooted crops list, 184
shears, for turning over beds, 106
sheep, for grazing cover crops, 45, 153
shovels
for forming raised beds, 55
for measuring soil compaction, 39
shredding of leaves, 82
sickle bar mowers, for cover crop termination, 158
side dressing of crops, 161, *161*
silage tarps
for long-season occultation, 42–43, *43*
potential negative effects of, 26–27
for soil temperature management, 163–64, *164*
for terminating cover crops, *96*, 153–55, *154*, *155*
for turning over beds, 96, *96*, 97–99
Simple Inoculating Compost Recipe, 61–64, 138, 139
Singing Frogs Farm, 70
single leader tomato plants, *231*, 233, 235, 236
site selection, 33–35
slips, sweet potato, 220–21, *221*
slope, site selection considerations, 34–35
slug control, 73
Small Axe Farm, 55
Smarter by Nature, 76
Smith, Jared, *55*, 104, 105–6, 172
sod, in pathways, 122–24, *123*
softneck vs. hardneck garlic, 205–6
See also garlic
soil aggregates
role in carbon sequestration, 13
role in soil stability, 23, 24
in soil organic matter, 14–15
soil amendments
fertility management principles, 143–48
incorporating amendments without tillage, 159–162, *160*, *161*
using results from soil tests, 40
when turning over beds, 95, *95*
soil biology, 140–41, *140*
soil blocks
for beets, 227
growing transplants in, 167–68
for lettuce, 214
Soil Fertility: A Guide to Organic and Inorganic Soil Amendments (Parnes), 101
soil food web, 140–41
Soil Food Web School, 141
soil health, importance of, 1–2, 6
See also specific practices
soil microorganisms
compacted soil effects on, 25–26
indigenous microorganism (IMO) collection, 138–39, *138*, *139*
inoculating composts for, 60–61
nitrogen-fixing bacteria, 146–47
nourishment by root exudates, 12, *21*
propagating, 143
role in photosynthesis, 19–20
in the soil food web, 141
solarization effects on, 111–12
symbiotic relationship with plants, 11, 12–13
testing for, 141
soil mix, 166–67, *167*
soil organic matter
from cover crops, *14*
defined, 14
determining percentage of, 15, 134–35
in fertility management, 134–35, *134*, 140
formation of, 14–15
mineralization of, 147–48
pores from, 135, *136*
tillage-related loss of, 24
soil organisms
role in photosynthesis, 15, 19–20, *19*
role in soil stability, 23
in soil food web, *21*
See also soil microorganisms
soil permeability, 137, 140, 142–43
soil probes, *37*, *134*
soil structure, 135–37, *136*, 142–43
soil temperature, planting determined by, 29
soil tests
determining percentage of organic matter by, 15, 134–35
lead contamination concerns, 35
microbial, 141
for nutritional status confirmation, 20
when starting from scratch, 35–37, *36*

soil texture, for seed germination, 162–63
soil types
  defined, 137
  site selection considerations, 35
solarization
  crimp and solarize method of cover crop termination, 156–57, *156*, *157*
  for turning over beds, 111–12
sorghum sudangrass
  cover crop use and termination, 158, 243
  for mulch-in-place living pathways, 126
sowing seeds. *See* seeding
soybeans (*Glycine max*)
  cover crop use and termination, 245
  relay cropping with wheat, 187
spacing of crops. *See* crop spacing
spinach
  critical period of competition, 252
  interplanting strategies, 180, 252
spin trimmers. *See* weed whackers
split beds, *178*, 180
spoiled hay, 75, 77
  See also hay mulch
spraying, of soil amendments, 160–61
squash, summer
  critical period and suggested pairings, 252
  relay cropping with, 188
squash, winter
  critical period and suggested pairings, 252–53
  with perennial cover crops, *150*
  in the Three Sisters combination, 174, 184
stale seedbed technique
  for arugula, 201
  for carrots, 196, 197
  overview, 89
standardizing bed width and length, 50–51, *51*, 53–55
Stand-up 35 Soil Blockers, 214, 227
staples, yard, 85
steam termination of cover crops, 159
stirrup hoes
  for cultivation, 47
  for pathway weed control, 127, 128
  for turning over beds, 110
Stoltzfus, Elam, 218
stomata, 11, 18–19
straw mulch, 72–74, 76, *76*
strip tillage systems, 77, 124
Stropharia rugosoannulata (wine cap mushroom), 115, *115*
subsoilers, appropriate use of, 27, 41
succession planting. *See* turning over beds
suckers, of tomatoes, 235–36, *236*
sunflower (*Helianthus annuus*), cover crop use and termination, 158, 243
sunlight
  mitigating excessive, 15, 16–17
  role in photosynthesis, *10*, 11, 15, 16–17, *16*
  seed germination needs, 162
  site selection considerations, 34
sunn hemp (*Crotalaria juncea*)
  cover crop use and termination, 158, 245
  for mulch-in-place living pathways, 126
Sustainable Agriculture Research and Education (SARE), 119
sweet alyssum, interplanting with, 177, *177*, 232
sweet clover (*Melilotus officinalis*, *M. albus*), cover crop use and termination, 245
sweet corn, interplanting strategies, 172, 179
sweet potatoes
  critical period and suggested pairings, 253
  in-depth profile of, 220–25, *221*, *222*, *224*
  dry farming of, 133, *133*
sweet potato greens, 225
synthetic mulches, 85–86, *85*, 119–120

T

Takemura, Yoko, 76, 161
tarping
  in the never-till approach, 42–44, *43*
  in the no-till market farm, *x*
  for soil temperature management, 163–64, *164*
  when transitioning to no-till, 45, *46*
  See also occultation; silage tarps
Taylor, Angelique, 76
Taylor, Cheezy, *104*
teff grass (*Eragrostis tef*), cover crop use and termination, 243
temperature
  compost, 62, 63, 64
  seed germination needs, 162, 163–64, 166
  transplanting considerations, 29, 169
termination of cover crops
  crimp and solarize method, 156–57, *156*, *157*
  crimp and tarp method, 153–55, *154*, *155*
  crimp/mow and mulch method, 157
  in fertility management, 150–59, *152*, *154*, *155*, *156*, *157*
  listing of crops, 241–46

termination of cover crops (*continued*)
mulch-in-place living pathways, 125–26
roller-crimping method, 151–52, *152*
seasonal considerations, 88
sickle bar mowers for, 158
silage tarps for, *96*, 153–55, *154*, *155*
steam, heat, and flame methods, 159
winter-killing, 88, 158
termination of crops
solarization for, 111–12
when turning over beds, 93, *93*, 94*t*, 110
See also occultation
terracing
at Sage Hill Ranch Gardnes, *48*
site selection considerations, 34–35
Three Sisters combination, 174, 184
Tierra Vida Farm, 45
tillage
animals for, 44–45, *44*
appropriate use of, 40
author's experience with, 89–90
definitions of, 22–23
soil degradation from, 23–27
tools for, 26–27
tillage radishes (*Raphanus sativus*)
for compacted soil, 42, 143
cover crop use and termination, 246
tillers, appropriate use of, 26
Tillie (electric tool), 55
tilthers, 160, *160*, 163
Tilth Soil, 167
time savings benefits of no-till practices, 30
tine rakes, 159
tomatoes
cherry tomatoes, 230–39, *231*, *234–37*
critical period of competition, 253
interplanting strategies, 172, *173*, 177, *177*, 179, 253
relay cropping with, 188, 190
root complementarity with lettuce, 184
tools and equipment
bed rakes, 159
bed rollers, 155, 162–63
bed width considerations, 47, 50
broadforks, 3–4, 26, 41, 46, *118*
bush hog mowers, 110
compost spreaders, 55, 70
for compost teas and extracts, *145*
crimping alternatives, 154–55, *154*, *155*
cultivating shoe attachments, 128
EarthWay seeders, 161, *161*, 197, 202, 227–28, 230
edgers, *123*, 124
flail mower collectors, 77, 109–10
flail mowers, 82, 87, 107–8, *108*, 109, *109*, 122
for forming raised beds, 55
Gridder tool, 207
harrows, 154–55, *155*, 163
hoes, 47, 106, 110–11, 127, 128
Hozon siphons, 161
Jang seeders, 153, 197, 202, *203*, 227–28
knives, 106–7, *107*
Landzie Compost and Peat Moss Spreaders, *83*, 84, 196, 221
list of manufacturers and distributors, 257
manure spreaders, 77
pinpoint seeders, 202
power harrows, 154–55, *155*, 163
precision depth rollers, 160, 163
ProtekNet, 205, 226
push mowers, 108–9
Quick-cut Greens Harvesters, 204, *204*
rakes, 108, 159, 163
roller crimpers, 53, 86, *88*, 151–52, *152*
rotary plows, 47, 117
seeders, 88, 153, 161, *161*, 197, 202, *203*, 227, 229
seed starting, 166–68, *167*
shears, 106
shovels, 39, 55
sickle bar mowers, 158
soil probes, 37, *134*
tilthers, 160, *160*, 163
walk-behind tractors, 30, 47, 55, 70, 107, *108*, 110
WeedGuardPlus, 78, 79, 101, 117
weed whackers, 102, *103*, 105, 122, 124, 157–58
wheelbarrows, 69
T-post trellises, 235
tractors, walk-behind. *See* walk-behind tractors
training, of cherry tomatoes, 233–36, *234*, *235*, *236*
transplants
beets, 226, 228
cherry tomatoes, 233
critical period of competition benefits of, 182
hardening off, 168–69
lettuce, 213, 214
nourishing with compost tea or extracts, 20, 169
planting of, 76, 77, 169
principles of growing healthy, 165–69, *166*, *167*
soaking root balls in compost teas and extracts, 145

soil temperature influence on, 29
sweet potatoes, 223
trellising, cherry tomatoes, 233–36, *234*, *235*, *236*
trialing of new varieties, 194
triticale (*Triticum aestivum* × *triticosacale*), cover crop use and termination, 243
Trump, Chris, 139
Turner Farm, 119
turning over beds, 89–112
for cherry tomatoes, 238
crop termination methods, 93, 94*t*
hand tool methods, 102–7, *103*, *106*, *107*
hoes for, 110–11
maintaining soil health during, 91–96, *92*, *93*, *95*
mowing methods, 107–10, *108*, *109*
number of times per growing season, *39*
occultation methods, 96–102, *96*, *100*, *102*
overview, 89–91, *90*
solarization for, 111–12
turnips (*Brassica rapa*)
cover crop use and termination, 246
critical period and suggested pairings, 253
relay cropping with, 188, 190
Twin Oaks, 190
Tyler, Ray, 99, 256

U

Université Laval, 116
unmulched pathways, 127–28, *127*
unused space, interplanting in, 172
Urbavore Urban Farm, *76*
US Composting Council, 68

V

vermicast (vermicompost), 61
Vermont Compost Company, 167
vernalization periods, 212
vetch (*Vicia* spp.)
cover crop use and termination, 245
for mulch-in-place living pathways, 126
Vivosun air pumps, *145*
vole control, 73

W

waffle gardening (Latdekwi:we), 54, *54*
walk-behind tractors
attachments for, 70
balers for, 110
flail mowers used with, 30, 107, *108*
for forming raised beds, 47, 55
water
for carrot beds, 195
in fertility management, 132–34, *133*, 142
orientation of beds and, 52
role in photosynthesis, *10*, 11, 15, 18–19
seed germination needs, 162, 168
for transplants, 169
See also irrigation
Web Soil Survey website, 137
WeedGuardPlus, 78, *79*, 101, 117
weeds
arugula, 201
beets, 226–27
carrots, 196–97
cherry tomatoes, 232
considerations when transitioning to no-till, 45–46, *46*
critical period of competition, 181–82, *181*
decreased numbers of, with no-till, 28–29
garlic, 206
from hay mulch, 74–75, *75*
landscape fabric for control of, 99
lettuce, 214
long-season occultation for control of, 42–43, *43*
mulching for control of, 28–29, *28*
no-mulch methods, 170–71, *170*
path management, 116, *123*, 124, 127–28, *127*
previous crops as, 93
removing when turning over beds, 108
site selection considerations, 35
from straw mulch, 73–74
sweet potatoes, 222–23
tillage effects on, 24–25
weed whackers
for cover crop termination, 157–58
for living pathways, 122, 124
when turning over beds, 102, *103*, 105
West Africa, zai pits system, 54, *54*
wheat (*Triticum aestivum*)
cover crop use and termination, 243
for mulch-in-place living pathways, 126
relay cropping with soybeans, 187
wheelbarrows, for applying compost, 69
wheel hoes
for pathway weed control, 128
for removing stubble, 106
for turning over beds, 110–11
white clover, in living pathways, *121*

width of pathways, 50, 120, 122
width of permanent beds, 47–50, *49*
Wild Hope Farm, 153
wildlife presence, *32*, *32*
windy areas
  orientation of beds in, 52
  securing silage tarps in, 97–98
wine cap mushrooms (*Stropharia rugosoannulata*), 115, *115*
Winstrip trays, 167–68
winter-killing
  for cover crop termination, 88, 158
  mulch-in-place living pathways, 125
winter squash, *150*, 252–53
wire trellises, for cherry tomatoes, 233–35, *234*
wood chips
  for mulch, 80–82, *81*
  on pathways, 80, 113–17, *114*, *115*, *116*
Woodleaf Farm, 77, 124
wool, as pathway mulch, 119
worms
  affinity for cardboard, 118
  as indicator of living soil, *19*, *36*, *136*
  vermicast from, 61
woven plastic silage tarps, 97, 98–99
  See also silage tarps

X

Xerces Society, 187

Y

yard staples, 85
yields
  arugula, 202, 204
  beets, 228–29
  carrots, 197, 199
  cherry tomatoes, 236, 238
  interplanting strategies, 179
  lettuce, 216–17
  sweet potatoes, 223–24
YYJ-24 rollers, 202

Z

zai pits system (West Africa), 54, *54*
zucchini
  critical period and suggested pairings, 252
  relay cropping with, 188, *189*
Zuni (A:shiwi) people, 54

About the Author

Cassie Lopez

JESSE FROST, aka Farmer Jesse, is a certified organic market gardener, freelance journalist, and the host of *The No-Till Market Garden Podcast*. He is also a cofounder of No-Till Growers (notillgrowers.com), where he helps collect the best and latest no-till insights from growers in the United States, Canada, the U.K., Australia, New Zealand, and Europe. He and his wife, Hannah Crabtree, practice no-till farming at Rough Draft Farmstead in central Kentucky.